NEURAL NETWORKS AND FUZZY LOGIC

Dr. R.P. Das

B.Tech. (Hons), Ph.D.
Principal
Pydah College of Engg., Visakhapatnam

L. Sreedhar

B.Tech. M.Tech, Head of R&D
Associated Professor
Pydah College of Engg. & Tech., Vishakhapatnam

S.K. KATARIA & SONS®

Publisher of Engineering Books
4885/109, Prakash Mahal, Dr. Subhash Bhargav Lane,
Daryaganj, New Delhi-110002 (India)
Phone: +91-11-43551243 Mobile: +91-9871775858, 7678425209
e-mail: katariabook@gmail.com, katariabooks@yahoo.com
Website: www.skkatariaandsons.com
(e-books, subscriptions and bulk buying also available on our website)

Published by:

S.K. Kataria & Sons®

4885/109, Prakash Mahal, Dr. Subhash Bhargav Lane,
Daryaganj, New Delhi-110002 (India)
Phone: +91-11-43551243 Mobile: +91-9871775858, 7678425209
e-mail: katariabook@gmail.com, katariabooks@yahoo.com
Website: www.skkatariaandsons.com

First Edition: 2012, Reprint: 2026

Laser Type Setting by:

SARA ASSIGNMENTS, Patparganj, Delhi-110092.

PREFACE

Neural networks and Fuzzy Logic are essentially step towards solving problems in "human" way. Neural Networks are inspired by the function of biological neurons. Assuming a certain pattern of behavior of neurons in the brain and nervous system, electronic networks can be built up for solving problems the way a human being would like to do. Many complex problems such as load forecasting pattern recognition and process optimization can be tackled by such networks. It was originally thought by Buddhist philosophers that we have "to make use of mind to think conceptually by leaving the substance and attaching self to form". The way human mind function is emulated by various artificial neural networks such as perceptron and associative memories.

Fuzzy Logic, on the other hand, tells us that all are not white or black – there is a grey area. Every event can have a membership function based on probability between 0 and 1. A city can have "the best of times and the worst of times" as in Dickens' description of London in the Tale of Two cities. Many real situations like a driver pressing the brake basing on assessment of obstacle being near or very near are real life fuzzy situations. Process control has innumerable possibilities which can be best dealt by assuming a fuzzy bias.

Neuro-fuzzy or Fuzzy-neuro techniques again throw many possibilities of solving complex problems faced in various fields. It is admirable that Electrical Engineers (EEE) and Computer Science Engineers are being imparted a course on the fundamentals of NNFL as per the syllabus of JNTU (K).

As teachers of the subject, we found lot of difficulties in advising students which books they should follow to cover the entire syllabus. Most of the books are written by authorities who have stressed on their areas of choice. As a result a student needs to study various portions of 8 to 10 books to meet the requirements. Moreover, most of the books available are almost blank on the models of neurons which are aptly selected by the syllabus makers. Without comprehending these, a student may become an Alice in the Wonderland.

It is our earnest endeavor to meet the requirements of students of 3rd year and Final year B.Tech of JNTU (K). No originality can be claimed except in presentation of matter in individual units. To explain multi category perceptrons, we have gone through classification (not in the syllabus as such). Many examples have been "chosen" from those of the standard authors – it is only hoped that the selection does justice to the topics. For instance, the problem on back propagation by one of the reputed authors is a unique presentation of such cases. The best problems of the individual authors have been presented in this book so that the student gets a "Geetasaar" out of the Upanishads of Zurada, Kosko, Haykin, Yagnanarayana among others. The objective of the effort will be met if the students and teachers do not run after many references for the purpose of completion of courses prescribed by JNTU (K).

Dr. R.P. Das

A

ACKNOWLEDGEMENT

The need for handy book on NNFL was the driving force behind this effort. To start with Mrs. Sangeetha, HOD, CSE Dept., must be thanked for creating the demand of an efficient neural and fuzzy subject teacher. The second input came from the students of EEE IV year, 2010 batch outs-Shyamala, Sanjeeivni, Vidhayadhari, Rohini who maintained excellent class notes.

Compilation of material was helped by Co-Author Mr. L. Sreedhar, Associate Prof. and I/C R&D unit. The students of final year (2011-batch passouts)-Sourav Puglia, Daniel and Tejaswani expedited the processing of the manuscript. Major credit of editing goes to Mrs. G. Nagamani, Asst Prof., CSE Dept.

The students of CSE for whom AINN course was prescribed did help to some extent in modifying the tenor of the chapters. An ECE student, P. Kiran Kumar Reddy has been responsible for the final touch up and checking.

At various stages my children Vighnesh and Gargie corrected the manuscript. My wife's patience has been outstanding during the two year period of preparation.

Dedicated
to

Pydah Krishna Prasad garu

Chairman
Pydah Educational Academy, Visakhapatnam

Without whose patronage and encouragement this effort would not have been completed.

C

CONTENTS

1

INTRODUCTION TO NEURAL NETWORKS

1.0 INTRODUCTION

Neural network can mean biological type as in human brain and artificial type as those based on working of the biological counterpart. In either case, the concepts of parallel distribution processing system consisting of computing elements having the ability to learn is the basic premise. A human brain develops with time, gaining experience. This involves the development of brain cells or neurons to adapt to the environment. Artificial Neural Networks (ANNs) are simplified copies of the biological counterpart existing in the central nervous system with the inspiration of computing done by the brain.

In a hierarchical order, the Biological Neural Networks (BNN) is: central nervous system → regional circuits' → local circuits' → Neurons. Again, the neurons operate in the order: Neurons → Dendrites → Neural microcircuits → Synapses → Molecules.

The most significantly common premise of ANN and BNN are the multiple inputs coming in with weighting pattern and adding up to an extent, which after crossing a barrier, activates the basic unit (neuron). The models of artificial neuron are justly reflections or imitations of the working of biological neuron as we understand so far. It must be stressed that the actual working of the human brain in total is far too complicated. Very simplistic working patterns can be simulated by ANN in some implementable form. Let us consider few diagrams to understand what neural networks mean.

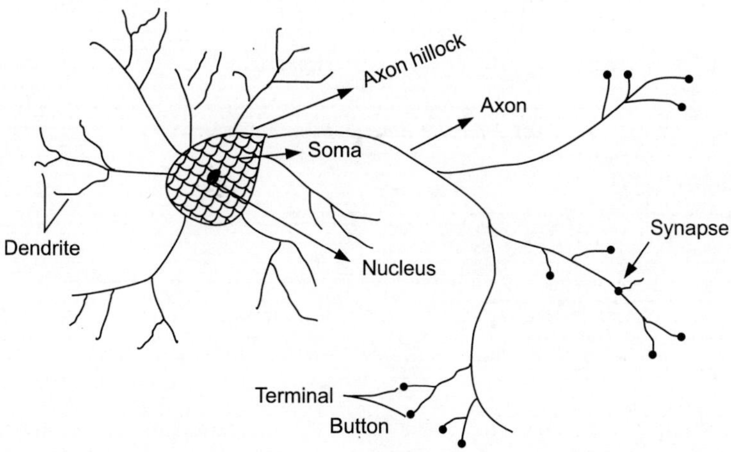

Fig. 1.1

A neuron is composed of cell body (soma), dendrites, Axons, and terminal synaptic buttons. The dendrites are tree like structures which pick up input signals from other neurons. Depending on the synaptic strength, the weighted sum of the input activates the neuron which fires a signal through the Axon once the sum exceeds a certain barrier of threshold. This rather simple concept is amenable to electronic implementation of artificial neurons.

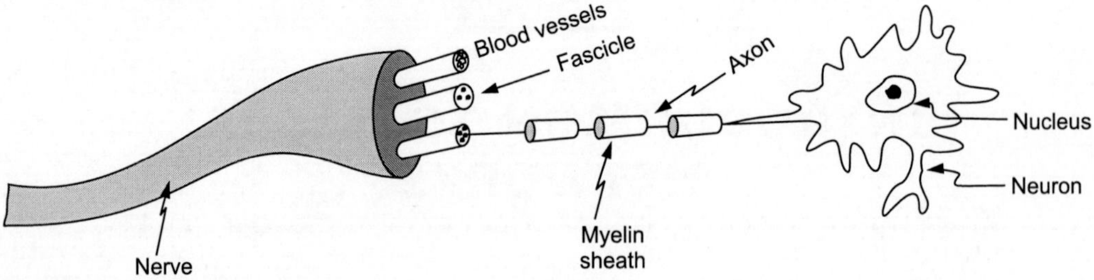

Fig. 1.2

Figure 1.3(a) shows an early conceptual implementation in the form of Perceptron by Rosenblatt. Figure 1.3(b) shows a schematic of an artificial neuron in an AND circuit. These two pictures essentially sum up the story. Following the brain's neuron, neural concepts can be developed to utilize the ideas in form of Logic gates.

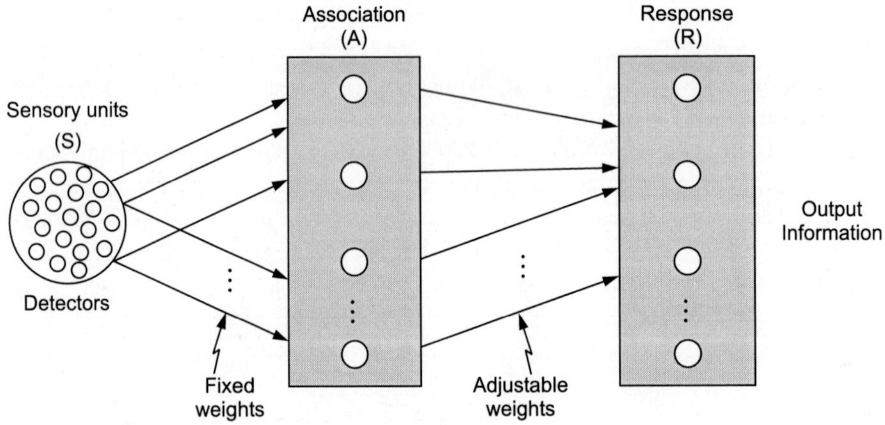

(a) *Model of Rosenblatt's Perception*

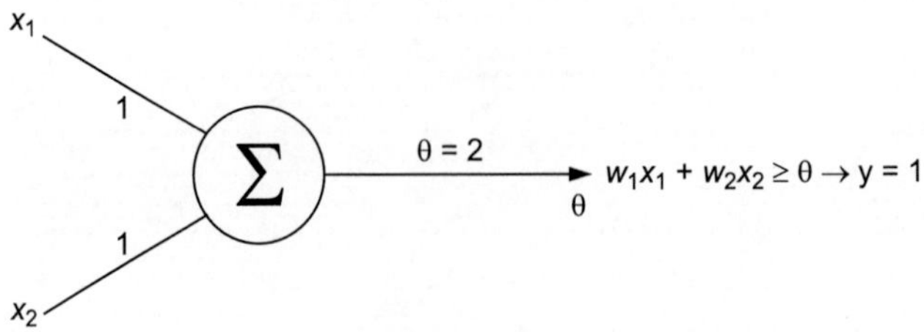

(b)

Fig. 1.3

1.0.1 Neuron

Neuron is a living cell which does not replicate itself but can develop more connections with other neurons (as a person learns more). There are nearly 10^{11} neurons in the brain and each neuron has 10^4 connections. The parts of anatomy of the cell are:

1. Cell body (soma) 2. Dendrites 3. Axons
2. Terminal buttons (synaptic terminals)

The dendrites pick up the input signals. The soma acts as the CPU. The Axon carries the output signal.

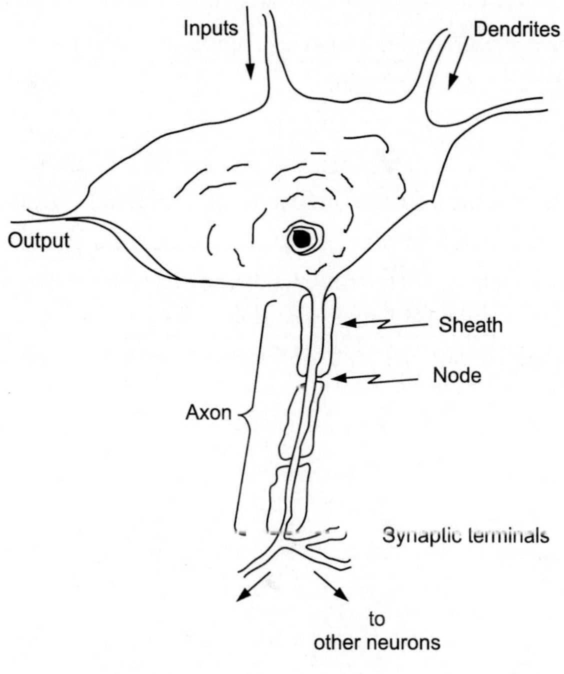

Fig. 1.4 *Structure of Neuron.*

1.0.2 Axon Structure

The central nervous system has glial cells which are of two types Astrocytes and Oligodendrocytes.

- Astrocytes are star like structures with long arms.
- Oligodendrocytes are shorter and have more numerous arms and have myelin sheath.

Another glial cell ("Schwann" cell) gives rise to and reduces the capacitance between the cell membrane and outside fluid for increasing the velocity of signal (120 m/s). The myelin sheath is segmented (each segment = few mm). There are gaps between segments. The nodes existing between gaps are called nodes of Ranvier.

1.0.3 Synapse

It is the point at which two or more neurons interconnect. The synaptic cleft consists of terminal buttons called 'synaptic terminals'.

1.0.4 Signal

The signal through Axon is transmitted through chemicals called 'potential' having low velocity. The chemicals can be modified meaning changing the velocity of the signal.

1.0.5 Basic Operation

- Dendrites receive synaptic inputs.
- The input signal current builds up an electric charge across cell membrane capacitance.
- At certain threshold a nerve pulse known as Action Potential (A.P.) is generated.
- AP is then propagated along the Axon.

1.0.6 Principle

- At rest a neuron is at some potential known as rest potential (of the order of 70 mV).
- Fluid inside the cell is rich in potassium (K^+) while the fluid outside is rich in sodium (Na^+).
- A metabolically driven pump in the cell membrane keeps the inside fluid rich in potassium and depleted in sodium (reverse for the outside fluid).
- The gradient of charged particles leads to potentials across the membrane.

$$\text{Reverse Potential } V_r = -\frac{KT}{q}\ln\left[\frac{N_{\text{int}}}{N_{ext}}\right] \quad ...(1.0.0)$$

- Each ion gradient produces at corresponding voltage.

1.0.7 Equivalent Circuit

When net current is zero and rest potential (V_0) appears $I = 0$,

$$I = 0 = \frac{V_K - V_0}{R_K} + \frac{V_{Na} - V_0}{R_{Na}} + \frac{V_{Cl} - V_0}{R_{Cl}} \quad ...(1.0.1)$$

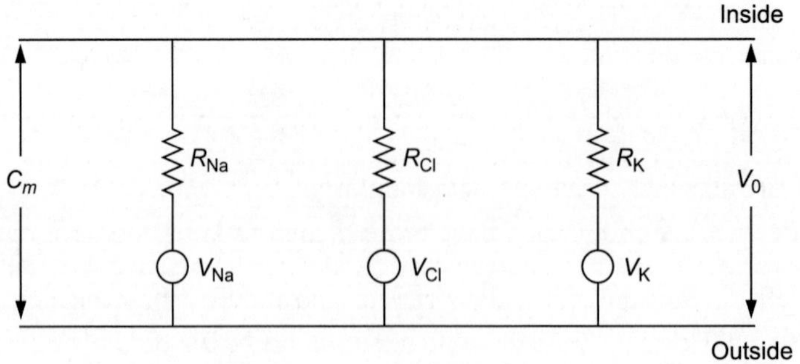

Fig. 1.5 *Equivalent Circuit*

$$V_0 = \frac{V_K R_{Na} + V_{Na} R_K}{R_K + R_{Na}} \text{ (neglecting Cl component)} \qquad ...(1.0.2)$$

The typical values are

$$C_m = 1\mu F/\text{Cm}^2$$

$$\rho_m = 5.6*10^9\ \Omega\ \text{Cm}$$

At rest a neuron is polarized to negative potential. Membrane being good insulator any application of excitatory signal depolarizes the membrane (decreases the negative potential).

An inhibitory signal hyper-polarizes (negative potential is increased) in biological network consisting of biological neuron, the input linkages are through synapses whose weights can be controlled by neurotransmitters in artificial neural networks (this action of weight change are simulated by numericals).

1.1 HUMAN SYSTEM AND COMPUTER SYSTEM

Parameter	*Human System (Biological)*	*Computer System (Artificial)*
1. Speed	These are slow and prompted by external stimulus. The order is millisecond.	These are faster. The cycle time depends on execution of program in the CPU.
2. Processing	There are massive parallel operations.	The operation is sequential.
3. Size and Complexity	The interconnections are 10^4 per neuron and there are 10^{11} neurons.	The corresponding number is much less such that complex recognition/processing is not possible.
4. Storage	Information is stored in interconnections. Information is adaptable as new information can be added by adjustment of interconnections without destroying the old information.	Information is stored in memory and there is a limitation in adding new information, as old information is destroyed.
5. Fault Tolerance	There is sufficient fault tolerance as information is distributed throughout the network.	Fault tolerance is less as information corrupted in the memory cannot be retrieved.
6. Control Mechanism	There is no central control for processing information. The neurons act on locally available information and transmit the output to the neurons which are interconnected.	There is a definite control system which monitors all the activities.

1.2 ORGANISATION OF BRAIN

1.2.1 Human Brain

The brain and the spinal cord form the central nervous system. The brain covers the top half of the head (shown in Fig. 1.6). It gets nutrient by a network of small blood cells between the skull and the brain. It weighs about 1.4 kg and it is divided into two hemispheres. Normally the brain to body ratio is of the order 1 : 50. The major parts are cerebrum, cerebellum and medulla oblongata.

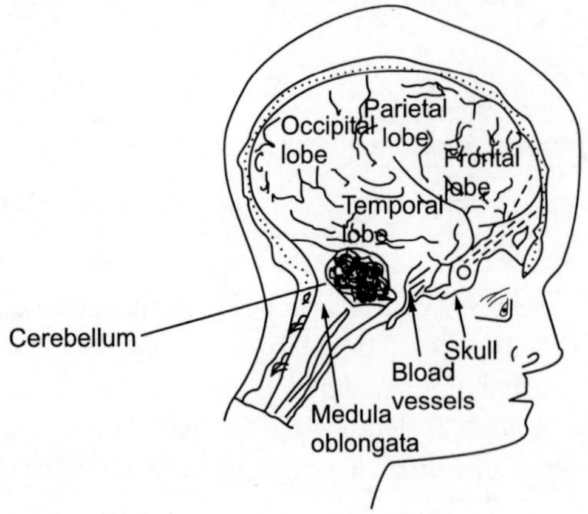

Fig. 1.6 *The Human Brain.*

1.2.2 Cerebrum

Cerebrum or the big brain has four different lobes: frontal, parietal, temporal and occipital. Its surface is heavily wrinkled so that more cells can be parked.

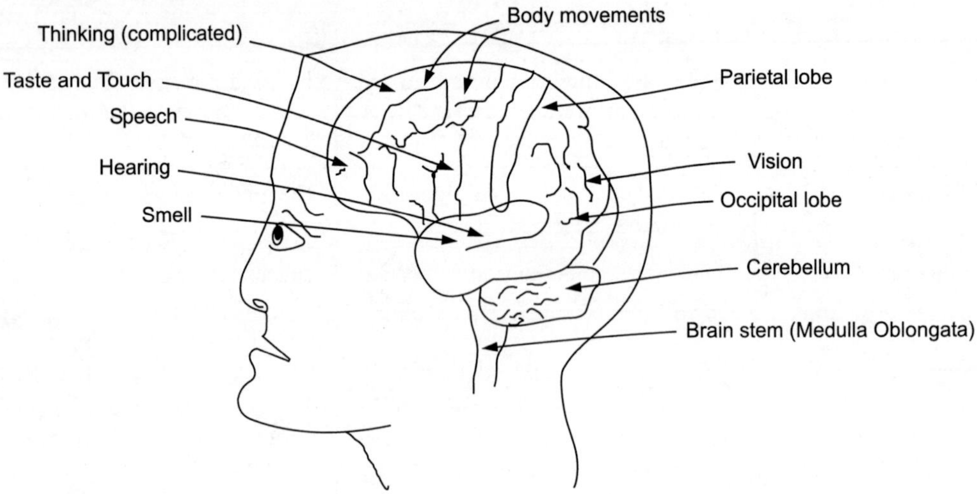

Fig. 1.7 *Location of some brain functions.*

1.2.3 Cerebellum

Cerebellum or the small brain is responsible for detecting and adjusting the body balance.

1.2.4 Medulla Oblongata

Medulla oblongata is the stem from which nerves carry any electrical impulses.

The brain contains nearly 10^{11} basic units called the neurons each of which is connected to 10^4 other neurons. Average weight of neuron is $(1.5) \times 10^{-9}$ gm. Those cells which perform input operations are called Afferent and those of which perform output operations are called Efferent. Other cells are used for interconnections.

The brain cells are similar to other body cells in the context of generation of bioelectric potential.

Two examples of human brain's behavior may be cited to explain the working of neural networks. When the finger touches an ordinary surface we perceive the touch and temperature of the surface. When the temperature of the surface increases beyond a certain limit the finger automatically detaches/retracts itself as if the brain has ordered it to do so. It is analogous to the condition of summation of inputs exceeding a certain barrier after which the neurons can fire and compel the removal of finger because of sense of higher temperature. This phenomenon can be thought as basic premise for building neural networks in which every neuron is supposed to get inputs of different synaptic weight values and the summation becomes worthy of a decision once it crosses a set value which in this case becomes the temperature tolerance capacity.

In another example we can think of the parameters a bride or a bridegroom evaluates, before taking a decision to marry. The inputs for judgement are height, color, education, background of the individual to be chosen. Each input has different weight-age like the synaptic weights. Once the summation of product of these inputs and the weight crosses a certain point, the selecting person decides on the choice much like the neurons firing.

In either case we can say that artificial neural ideas draw heavily from the way the human brain is supposed to operate as a whole or as a part of it in the form of the basic unit—the neuron.

1.2.5 Types and Dimension of Neuron

Inter-neurons are connected to other general neurons where as Motor neurons are connected to muscle fibers. Receptor neurons receive information from sensory organs (eye, ear).

This size of the cell body of a typical neuron is in the range 10 to 30 μm. Dendrite and Axons have diameter of the order of μm. The gap of the synaptic connection is about 200 nm. The total length of Axon varies from 1 mm to 1 m.

1.2.6 Cell Potentials

In the 18th century, the Galvani showed that there is potential difference across the membrane of a cell. Inside a negative charge and outside a positive charge exist with a potential difference of 70 to 90 mv (rest potential).

In the resting position the cell is polarized and the resting potential exists. If the cell is externally stimulated momentarily the outside of the cell acquires negative charge and inside a positive charge so that cell is depolarized.

Action potential can be compared to the output of one shot or monostable multivibrator which after triggering gives an output pulse of fixed width and returns to the previous state (repeated trigger pulse do not change the state).

After re-polarisation (returning to set potential) the cell does not respond immediately even if stimulus is applied; the gap is called refractory period.

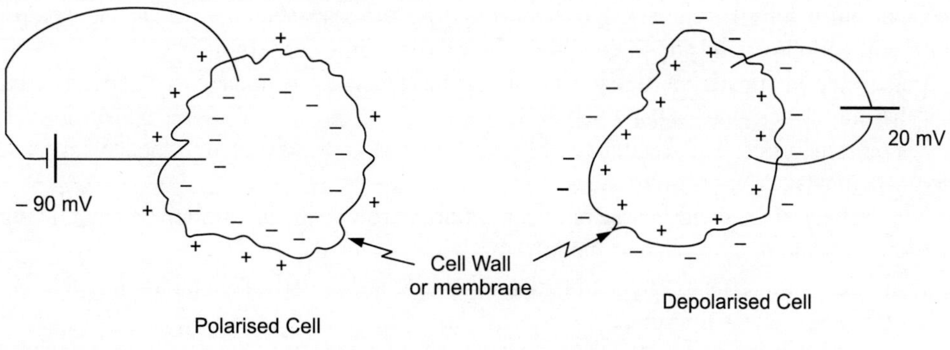

Fig. 1.8

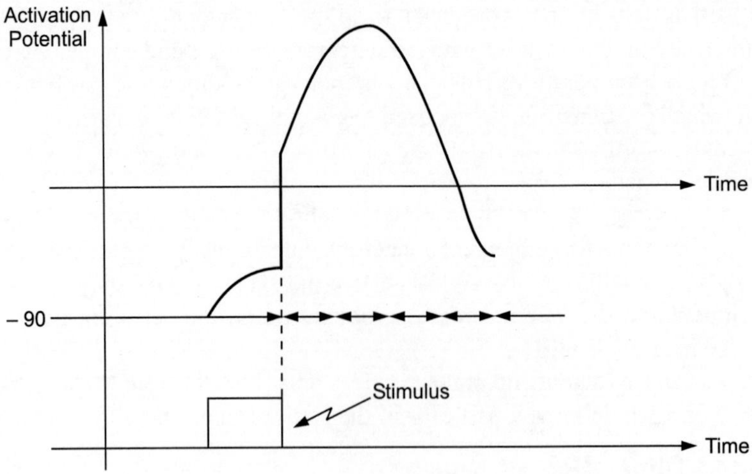

Fig. 1.9

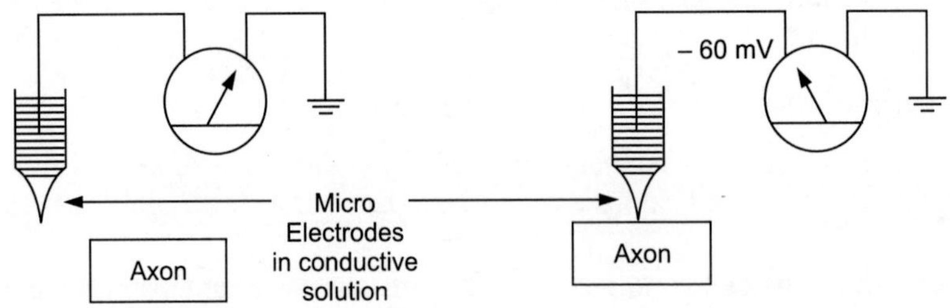

Fig. 1.10

1.3 BIOLOGICAL NEURON

A biological neuron or a nerve cell consists of synapses, dendrites, the cell body (SOMA), hillock and the AXON.A synapse is a bio chemical device which converts pre synaptic electric signal into chemical signal and then back to post synaptic electric signal.

The input pulse train has its amplitude modified by the parameters stored in the synapses. The nature of modification depends on the type of synapse (excitatory or inhibitory).

The post synaptic signals are transferred along the dendrites to other nerve cells. The cell body generates the output neuronal signal such as spike which is transferred along the Axon to synaptic terminals of other neurons. The frequency of firing a neuron is proportional to total synaptic activities and is controlled by neuron transmitters.

The morphology of the dendrite tree plays an important role in the integration of synaptic inputs and influences the way the neuron possess the information.

The resulting input suffers voltage attenuation so that about half of the total charge injected into distal dendritic sites reaches the soma.

The strengths of the charges received by the neuron on its dendrites are added together through a non-linear process of spatial and temporal summation, at the axon hillock region.

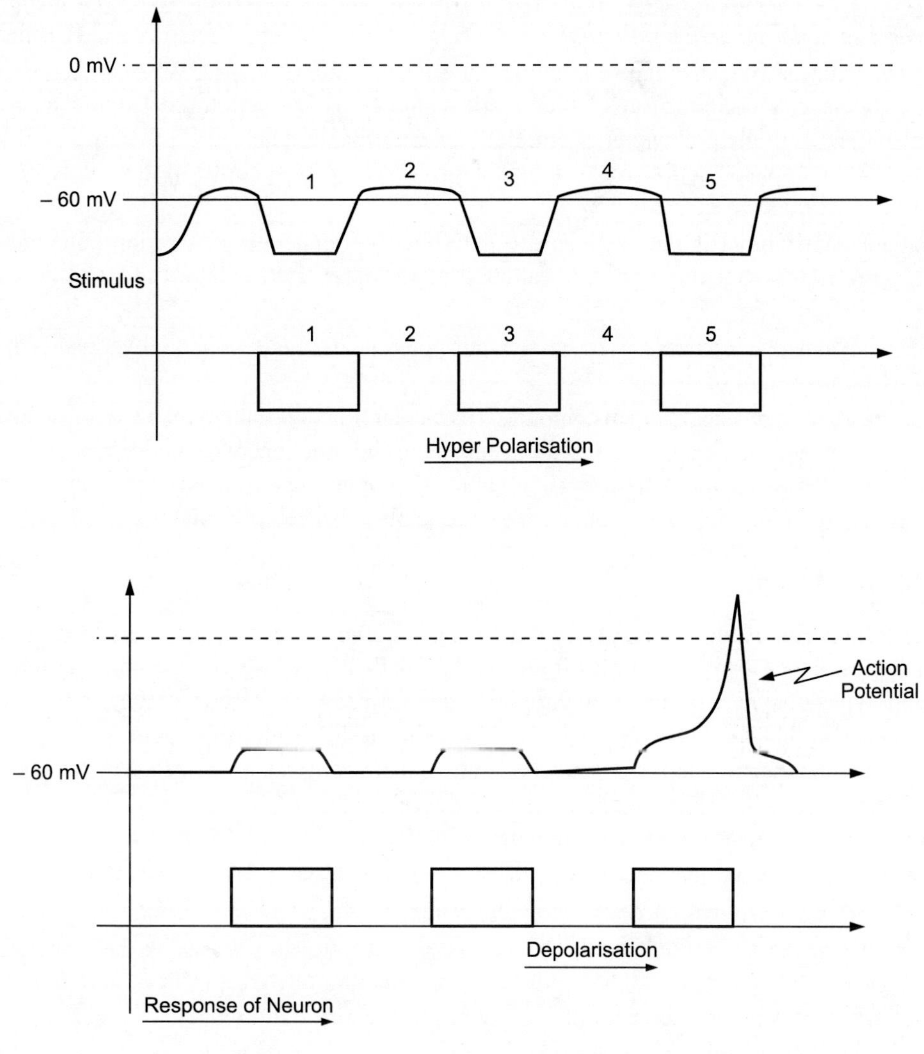

Fig. 1.11

The two parameters critical for spatio temporal integration of the synaptic inputs are: λ–the space constant, and τ_m membrane time constant.

The input from a dendritic site is characterized by a propagation delay. In case of slowly varying inputs the propagation delay is dictated by τ_m and for first synaptic inputs, it influences the τ_m.

The part of the soma which performs the important non-linear processing of the input is the AXON hillock. If the total input produces depolarization to the limit of the threshold (around –60 mv) the Axon hillock fires the action potential. The output signal is transmitted down the AXON and is delivered to other neurons.

After the firing, which is referred to as all/none response; there is a period of refractory behavior. During this period, no further action potential is generated.

"The minimal distance between two spikes defines absolute refractory period of the neurons. The above refractory period is followed by a phase of relative refractiveness during which it is difficult but not impossible to generate action potential" – Gerstner and Kilstner (2002)

The most common type of synapse in human brain is chemical synapse. When an action potential arrives at a synapse, a complex set of biochemical processes are generated leading to release of neuro-transmitters from the pre-synaptic terminal into the synaptic gap.

The voltage response of the pre-synaptic neuron to a post-synaptic potential is referred as post-synaptic potentials.

A single synaptic input is not sufficient to determine the generation of action potential. This is usually triggered as a consequence of non-linear introduction of several excitatory and inhibitory synaptic inputs.

At the end of synaptic terminals, the post-synaptic potential decays exponentially towards the resting value with a time constant of τ_m.

Strictly speaking, a physical system computes a particular function if there is an appropriate, mapping between physical states of the system, and the elements of the function. We can describe the stochastic behavior of a neuron which maps an activation value '*V*' a binary response which is a spike in 25% of the cases and no spike in 25% of the cases with a probability distribution function.

1.3.1 Other Methods

The model proposed by Hodgkin and Huxley in 1952 gives a very accurate and quantitative description of a physiological data. However complex frame works like this model which has numerous ion channels and different types of synapses are difficult to construct for analysis. As conception drawback for this model, a numeral complexity, involving number of non-linear differential equations.

Models of Hodgkin and Huxley can be applied to explain complex cases which mean neurons with any special structure and interconnections. It is known that the architecture of dendrite tree play an important role on non-linear functioning of neurons.

For construction of neuronal models considering all complexities, it is required to divide the neurons into number of finite iso-potential connected compartments.

Compartmental models are implemented by biologically realistic simulators called Genesis and Neuron.

A biological solution is to have the model of spiking nature of the neurons retaining the essential elements of behavior while trying to simplify the complex dendritic influences.

Several simplified neural models proposed have been:

Leaky integrated and fire, Resonant and fire.

1.4 BIOLOGICAL AND ARTIFICIAL NEURON MODELS

1.4.1 Neural Operation

Activation function $x(t)$ and signal $S(x(t))$ involves electrical pulses of potential difference and temporal summation. Activations involve small membrane pulses. Signals involve large Axonal pulses or action potential.

Potential difference in electrical charge occurs between the inside and outside of neurons surface membrane. Neuro transmitter chemical signals from other neurons through abutting synapses modulate membrane potential. Synapses transduces pulse coded electrical neuron signal with neuro transmitter signals. Membrane potential difference or pulses appear to accumulate at the membrane of Axon hillock where neuron converts its Axons or long branches.

If the potential difference pulses arriving at Axon hillock exceed a time varying threshold, Axon hillock membrane generates a signal pulse or action potential.

The large signal pulse propagates from Axon and its branches where Axonal insulators restore and amplify the signal as it propagates till it reaches a synaptic junction.

Synaptic process transducer the electric signal into chemical neuro transmitter signal which affects the potential difference across the membrane of the abutting post synaptic neuron.

At molecular level, ions drive changes in potential difference. Ionic flow changes the Axon hillock membrane potential by modifying the membrane conductance. Conductance change involves selectively opening and closing molecular channels Na^+ and K^+ flow across the membrane through these channels.

Na^+ and K^+ ions change their conductance, their ability to flow as they flow across the membrane by altering the molecular channels in complex way. Na^+ position are antagonistic (Na^+ excitatory, K^+ inhibitory).

Signal threshold summarizes the molecular channels through which the ions flow. Threshold effect is inversely proportional to the number of open molecular channels.

The more open molecular channels available for current flow, the lower the threshold that temporally summed potential difference pulses must exceed to emit a signal pulse.

Cl^+ chloride and other ions also affect neuronal signal transmission and reception. Real values $x(t)$ represents the membrane potential or voltage difference across the neurons surface membrane at time t– in the sense of temporal summation. Activation can be positive or negative can even be infinite.

$S(x)$ induced by x represent neurons firing frequency of action potential or pulses in sampling interval? Biological neuron is 10 to 30 ms: $S(x)$ should be non-negative frequency or signed signal may appear.

1.4.2 Simplified Model of Artificial Neuron

Threshold unit = Non-linear filter

$$\text{input} = \frac{\sum_{i=1}^{n} w_i x_i}{s} \quad \text{activation function} \qquad \ldots(1.4.1)$$

Step function

$$\phi(s) = 1 \text{ if } x \geq 0$$
$$= 0 \text{ if s} \leq 0 \qquad \ldots(1.4.2)$$

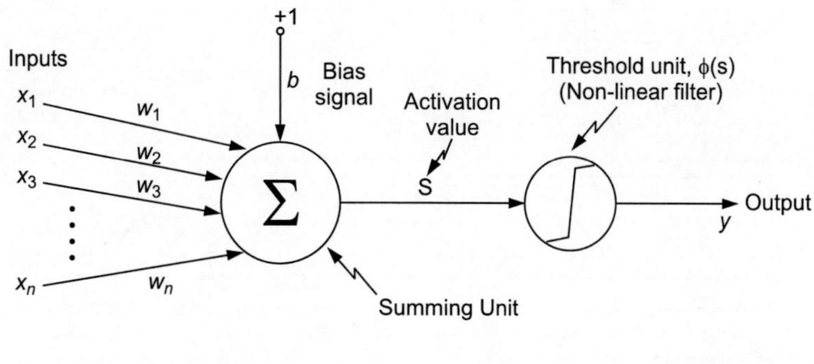

Fig. 1.12 *Simplified Artificial Neuron Model.*

1.4.3 Neural Behavior

Neurons behave as generator of functions. These transducer an activation $x(t)$ at t into a bounded output signal $S(x(t))$.

Sigmoidal curve of transduction is input output behavior of Op-Amps. Threshold signal function illustrates a non-differentiable signal function. In general signal functions are piece wise differentiable.

Threshold function gives rise to dashed lines. The family of logistic signal functions having C index approach asymptotically the threshold function as C increases to positive ∞.

S transduces positive activation x to unity signal negative activation to zero. A four neuron vector of activations (–6, 350, 49, –689) will become (0110). Sodium ions are excitatory. Potassium ions are inhibitory. Conductance change of membrane involves opening and closing of molecular channels. Sodium increases membrane potential where as potassium decreases input signals are passed onto cell body through synapse which may accelerate or retard. Acceleration or retardation is related to weights reflective synapse has a larger weight than a normal wave synapse.

Signal $S(x)$ induced by activation x represents neurons firing frequency of action potentials or pulses in sampling interval. In biological neurons the sampling interval may be in the range of 10 to 30 ms. Firing frequency equals the average number of pulses emitted in sampling interval.

1.5 HODGKIN-HUXLEY MODEL

The current understanding of neural excitability has been influenced by the research work HODGKIN and HUXLEY. In early fifties, AL Hodgkin and AF HUXLEY (along with B.Katz) developed mathematical model to explain the behavior of nerve cell.

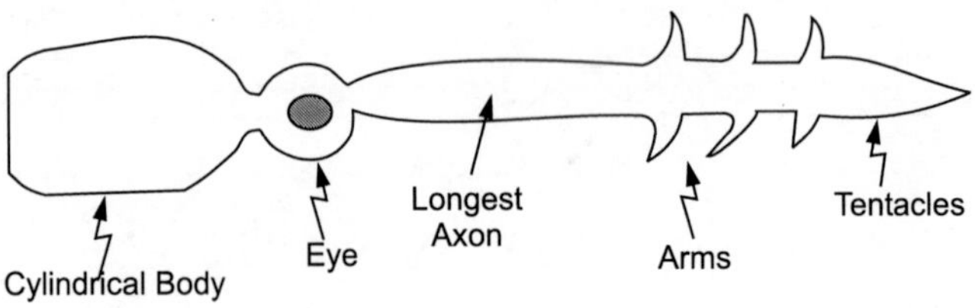

Fig. 1.13 *Neuronal Structure of the Squid.*

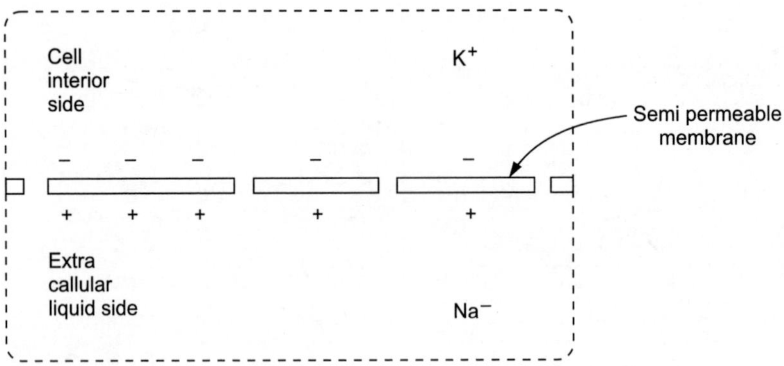

Fig. 1.14 *Cell schematic.*

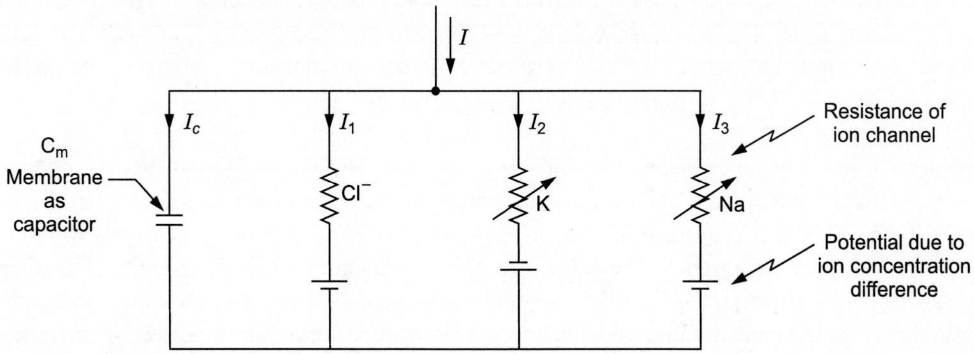

Fig. 1.15 *Cell Equivalent Electric Circuit Model.*

HODGKIN and HUXLEY experimented on giant AXON of squid (cephalopod mollusk shown in Fig. 1.14). Different types of ion current: Sodium, Potassium and leakage current which consists of Cl^- ions Specific voltage dependent ion channels, one for Sodium and another for potassium control flow of ions through cell membranes.

The leakage current takes care of other channel types which are implicit and negligible. The semi-permeable cell membrane separates the interior of cell from the extra cellular liquid.

If an input current $I(t)$ is injected into the cell, it may further add charge on the capacitor or leak through the channels in the membrane. Due to the active ion transport through the cell membrane the ion concentration in the cell is different from that the extra cellular liquid.

The Nernst potential generated by the difference in the ion concentration is represented by the battery. The conservation of an electric charge implies that the applied current may be split into I_c which charges the capacitor and I_k which passes through other ion channels.

Mathematically conservation of electric charge is

$$I(t) = I_c(t) + \sum_k I_k(t) \qquad ...(1.5.1)$$

where, I_c = charging current of capacitor

I_k = current of ion channel

and $$C_m = \frac{Q}{V}$$

Where Q = charge of electron

V = voltage across the membrane

By definition capacitance $C = Q/V$ where Q is the charge and V is the voltage across the capacitor. All the channels are characterized by the resistance (or conductance). The leakage channels are described by a voltage independent conductance g_L where as the conductance of other ion channels is voltage and time dependent.

If all channels are open, they transmit the currents with maximum conductance or respectively (normally some channels are blocked). The probability that a channel is open can be described by additional variables m, n and h. The combined action of m and h controls sodium channels, where as potassium channels are controlled by n.

Empirical equation is given by

$$\Sigma I_k = g_{Na} m^3 h(V - E_{Na}) + g_K n^4 (V - E_K) + g_{Cl}(V - E_{Cl}) \qquad ...(1.5.2)$$

The parameter E_{Na}, E_K, and E_{Cl} are called the reversal potentials. The values of these potentials are based on a voltage scale. The corrected values for E_{Na} is 50 mV and that of E_K is –77 mV. The table values of *m*, *n* and *h* as function of α* and β* these variables are also called as gating variables (also called rate constants). Reverse potential shifting by –65 mV gives E_{Na} = 115 mV, E_K = 12 mV E_{Cl} = 10.6 mV.

In order to interrupt the experimental data, Hodgkin and Huxley suggested that the conductance G_{Na} and G_k changed dynamically as function of membrane voltage (now it is known that the basis of their voltage dependence is related to the bio physical properties of membrane channels which control flow of ions across the membrane). These macroscopic conductances can be thought of as a result of combined effects of a large number of microscopic ion channels within the membrane. Each individual ion channel can have small number of physical gate switch to regulate the flow of ions; again an individual gate can be in one of two states permissive or non-permissive.

For a particular type of gates, a probability P_i can be defined for representing the probability of an individual gate being in the permissive state. This P_i can be thought of as the fraction of gates which are on the permissive state. Assuming first order kinetics we can write as,

$$\frac{dP_i}{dt} = \alpha_i(V)(1 - P_i) + \beta_i(V)P_i \qquad ...(1.5.3)$$

Where $\alpha_i(V)$ and $\beta_i(V)$ = voltage dependent constants (gating variables).

If the membrane potential V_m is fixed to V, then the fraction of gates in the permissible state will reach a steady value $\left(\frac{dP_i}{dt} = 0\right)$ as $t \to \infty$ given by

$$P_{i_{t\to\infty}} = \frac{\alpha_i(-V)}{\alpha_i(V) + \beta_i(V)} \qquad ...(1.5.4)$$

A simple exponential time constant is given by,

$$\tau_i = \frac{1}{\alpha_i(V) + \beta_i(V)} \qquad ...(1.5.5)$$

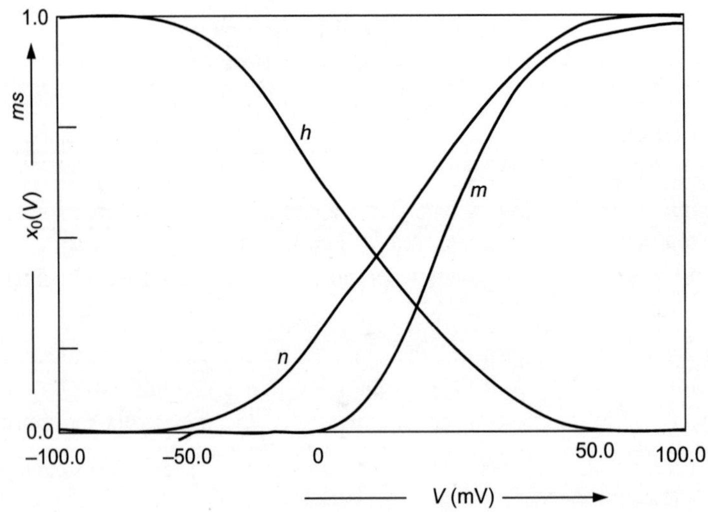

Fig. 1.16 *Equilibrium Function.*

When an individual channel is open (all gates are in the permissive state), there is a small fixed value added to the total conductance. Combined conductance of a large population of channels is proportional to the probability that the associated gates are in their permissive position. So, the conductance of channel as the constituent gates of type *I* is proportional to the product of individual gate probabilities P_i.

Hodgkin and Huxley modeled Na conductance using the gates of a type labelled *n* and another type *h*. *K* is modeled by 4 identical *n* gates. So Na conductance is proportional to m^4.

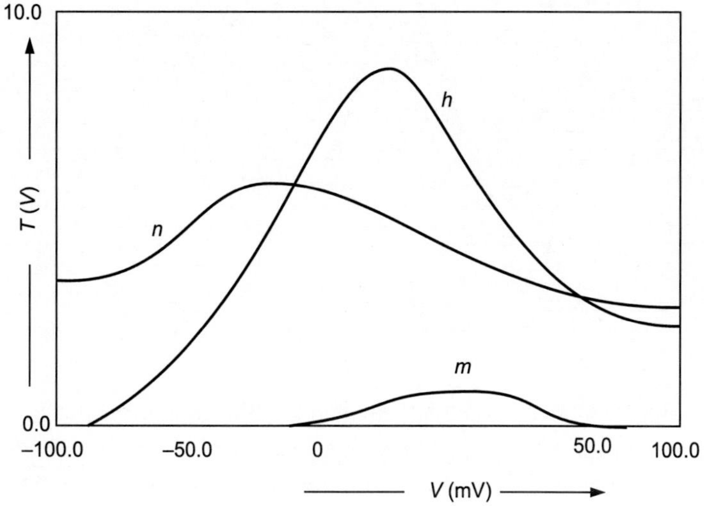

Fig. 1.17 *Time Constant.*

Equation Function χ_0	α	β
n	$\dfrac{(0.1-0.01V)}{\{\exp[1-0.1V]-1\}}$	$0.125\exp\left(-\dfrac{V}{80}\right)$
m	$\dfrac{2.5-0.1V}{\exp(2.5-0.1V)-1}$	$4\exp\left(-\dfrac{V}{18}\right)$
h	$0.07\exp\left(-\dfrac{V}{20}\right)$	$\dfrac{1}{\{\exp(3-0.18)V+1}$

$$\frac{dm}{dt} = \alpha_m(V)(1-m)-\beta_m(V)\,m$$

$$\frac{dh}{dt} = \alpha_h(V)(1-h)-\beta_h(V)\,h,$$

$$\frac{dn}{dt} = \alpha_n(V)(1-n)-\beta_n(V)\,n$$

the values of m and n increase with voltage, whereas the value of 'h' decreases. If some external input causes the membranes voltage to raise the conductance of sodium channels increases reflecting increase in 'm'. As a result, the positive, sodium ions flow into the cell and the membrane potential rises further. If this positive feedback is large and action potential can be generated.

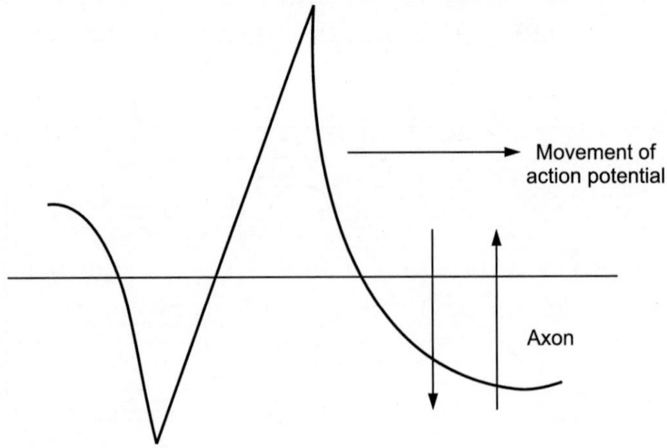

Fig. 1.18

These are parameters which are non mathematical but logical/empirical selected by Hodgkin and Huxley to feed the data of the Squid with the values already observed. x stands for m, n, or h. For fixed voltage V, the variable x (m, n, or h) approaches the values $x_o(V)$ with time constant of $T_x(V) = \tau$. The asymptotic values of $X_o(V)$ and time constant $T_x(V)$ can be written as:

$$\frac{dx}{dt} = -\frac{1}{T_x(V)}[x - x_0(V)] \quad \text{...(1.5.6)}$$

$$\frac{dm}{dt} = \dot{m} = \alpha_m(1-m) + \beta_m m \quad \text{...(1.5.7)}$$

$$x_0 = \frac{\alpha_x(V)}{\alpha_x(V) + \beta_x(V)} \quad \text{...(1.5.8)}$$

Flow of Action Potential

1. When an action potential occurs in membrane of the cell huge changes occur in permeability, the resistance value drop and large ionic current flows.
2. When a patch of Axon fires the action potential current causes other parts of the membrane to exceed the threshold value.
3. The action potential moves to new patches of membrane and travels down the Axon by igniting parts of the membrane.

1.5.1 Spike Generation

Refer to Fig. 1.19 on Action Potential (AP) The stimulation is by a strong current pulse (short) fixed before $i = 0$. In case of time $t > 0$. The value of,

$$\Delta V(t) = V(t) - V_{rest} \quad \text{...(1.5.9)}$$

Equation 1.5.9 shows a positive peak (AP) followed by a refractory period which the potential is below the resting potential.

The time course for $[V(t) - V_{rest}]$ is called KERNEL η.

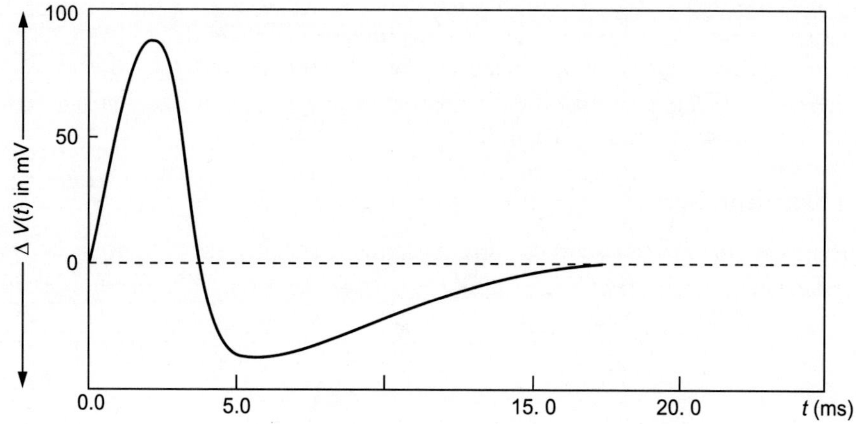

Fig. 1.19 *Action Potential.*

Refer to Fig. 1.20 which shows the threshold effect in the initiating of AP. A current pulse of one millisecond duration has been applied at t = 10 m sec. For current amplitude of 7 μA/cm^2, an AP with amplitude of 100 mV is initiated. If the stimulating current pulse is slightly weak, no action potential is emitted and a voltage remains below 10 mV.

At the high values of V, Sodium conductance is shut off due to h (Fig. 1.18). The time constant for 'h' is always larger than time constant for 'm'. So the variable 'h' which closes the channel reacts more slowly to the voltage increases then the variable 'm' which opens the channel. For a similar slow time scale the potassium current enters and lowers the potential. The overall effect this currents is a short AP, followed by a negative overshoot (refractory period Fig 1.19). The amplitude of spike is nearly 100 mV.

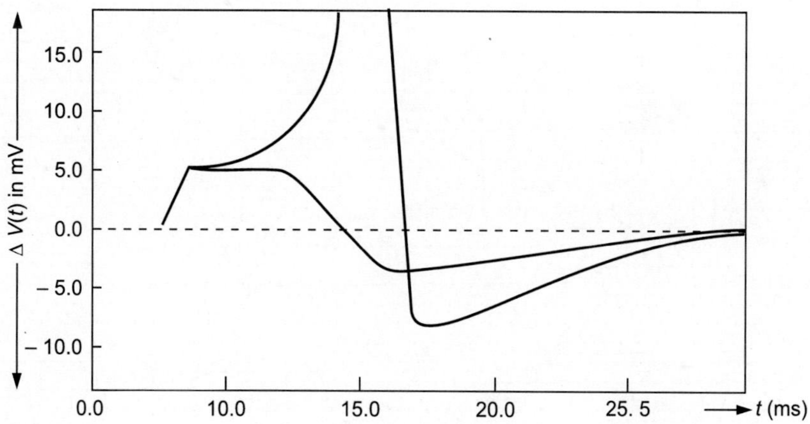

Fig. 1.20

In Fig. 1.19, the spike has been initiated by a short current pulse of 1 mS duration at $t < 0$. In case the amplitude of input situation is reduced below some critical value, the membrane potential comes base to the resting value without any large spike as in Fig. 1.20. So we can observe a threshold behaviour.

1.5.2 Mean Firing Rates and Gain Function

The basic equations of Hodgkin-Huxley can be studied for constant input $I(t) = I_0$ for $t > 0$. If the values of I_0 is larger than critical value I_θ (approximately 6 μA/cm^2) regular spiking is observed Fig. 1.21.

A firing rate $\nu = 1/T$ can be defined for T as interspike interval. Gain function is seen as the firing rate as a function of constant usual I_0 Fig. 1.22.

1.5.3 Step Current Input

Response of Hodgkin-Huxley modal to step current is quite important. Taking current $I(t) = I_1 + \Delta_1 H(t)$ where $H(t)$ is Heaviside step function ($H(t) = 0$ for $t \leq 0$ and $H(t) = 1$ for $t > 0$), the input is made to jump from I_1 to $I_2 = I_1 + \Delta I_1$ Fig. 1.23.

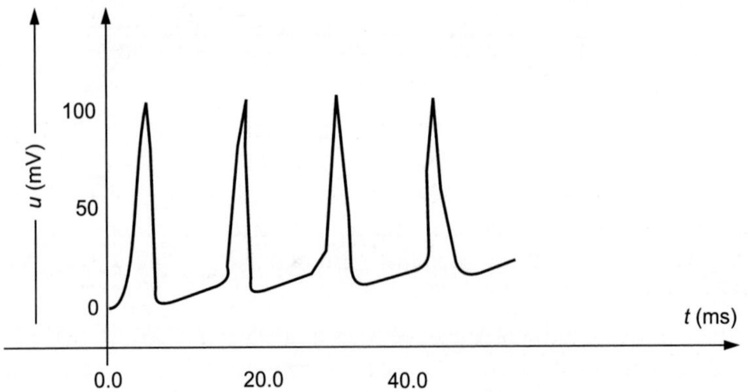

Fig. 1.21 *Spike Train for Constant Input I_0.*

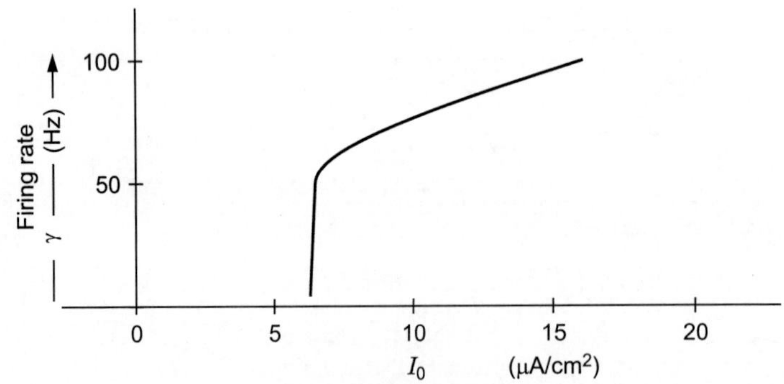

Fig. 1.22

Figs. 1.21, Fig. 1.22, Fig. 1.23 refer to various spike studies in the Hodgkin-Huxley model. If the current is larger than a critical value, there is regular spiking as in Fig. 1.21. The firing rate as a function of the constant input current is plotted in Fig. 1.22. When a step current in Fig. 1.23 appears, the results can be seen in Fig. 1.24 and to 1.25.

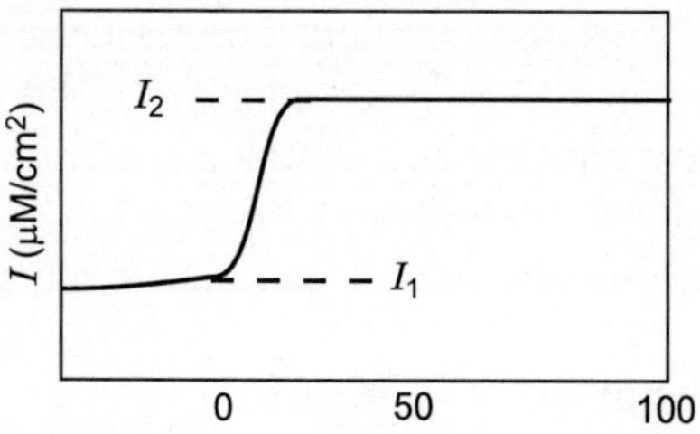

Fig. 1.23 *Step Input Current.*

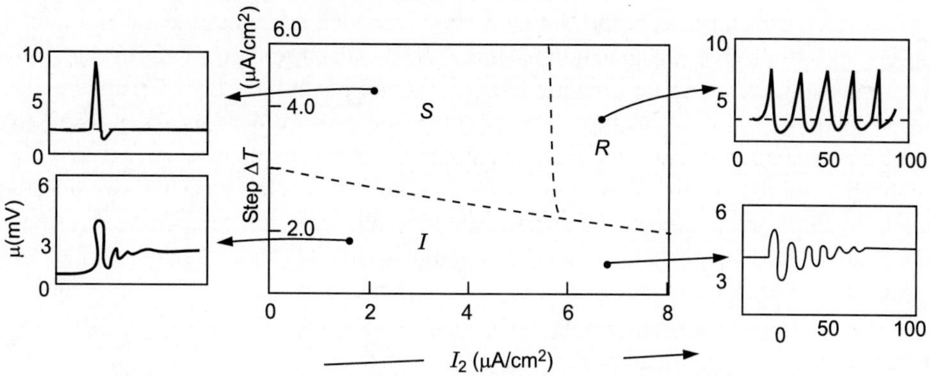

Fig. 1.24 *Response of the Hodgkin-Huxley Model to Step Current Input.*

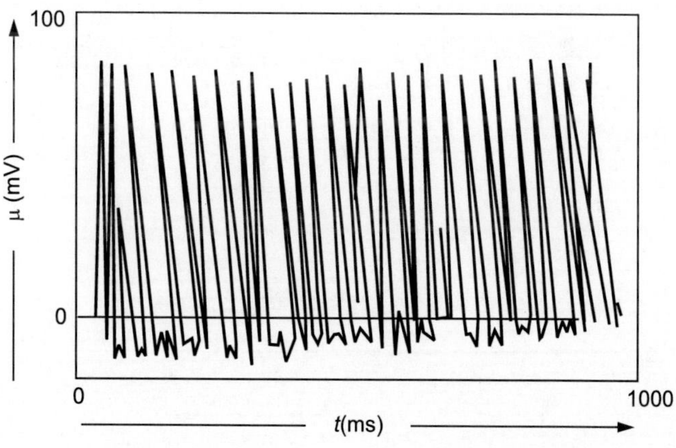

Fig. 1.25 *Spike Train for Time Dependent Case.*
(Driven by Time Dependent Input)

Referring to Fig. 1.24, it may be observed that large step ΔI facilitate spike initiation. Even for a target value of $I_2 = 0$ a spike is possible if the step size is large. This is case of inhibitory rebound in which a single spike is fixed if an inhibitory current $I_1 < 0$ is released. The region *S* stands for this effect of single spike.

The region *R* (Repetitive) shows repetitive firing for $I_2 > 6$ μA/cm^2 (triggering by large current steps is necessary).

No action potential is initiated in the region *I* (inactive). In conclusion, it is seen from this figure that there is no unique current threshold for spike initiation. This implies that trigger mechanism depends not only that current but also on the step size. Such observation makes the concept of threshold questionable (Kochetal 1995). Actually the transition is smooth though it appears like the threshold.

If the current between 6.9 and 7.0 μA/cm^2 can be tuned, a family of response amplitudes can be determined. For practical purposes, this transition is taken as the threshold.

1.5.4 Stimulation by Time-dependent Input

Fig. 1.25 shows a spike train of Hodgkin-Huxley model driven by a time dependent input current. The action potentials occur regularly.

Fig. 1.26 shows refractoriness of the Hodgkin-Huxley model. At 20 mS the model is stimulated by a short current pulse for triggering an action potential. A second current pulse of same amplitude at 25, 275, 308 32.5 ms strong enough for creating a second action potential.

A second spike is possible only after 15 ms after the first stimulation. It is also possible to trigger a second spike after a shorter internal if a stronger stimulation is given initially.

The voltage trajectory of this figure shows that the refractoriness has basically two different forms. Firstly due to hyperpolarisation spike after potential the voltage is reduced. This necessitates greater threshold for reaching firing threshold. Secondly as a large portion of channels remains after immediately after the spike, the resistance of the membrane is lowered in comparison with the situation at rest. Due to the depolarising effect of the stimulation, pulse decays faster after the spike in comparison with the situation after 10 ms.

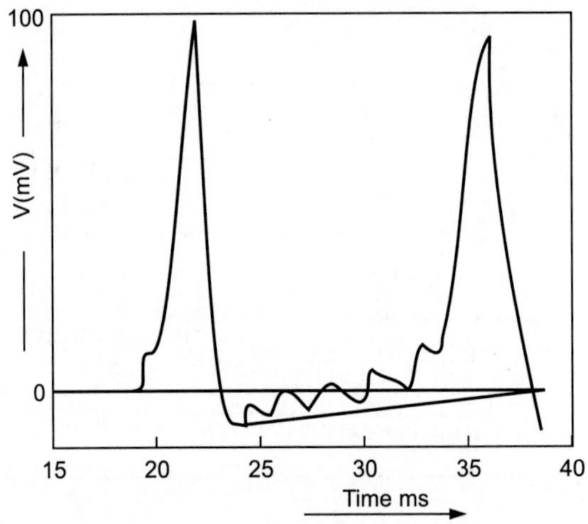

Fig. 1.26

1.6 INTEGRATED AND FIRE

1. In 1972, Bruce and Knight analyzed a simple integrator model of a single neuron on the basis of their study on a crab called Limulus.
2. By an approximation of the equation used by Hodgkin and Huxley for the Squid, it is possible to derive the equation of integration.
3. The model of Bruce and knight is also called integrate and fire model. An allied version is forgetful integrate and fire model.

4. Assume a noise free neuron containing an internal variable like membrane potential. Let it be called $V(t)$.
5. A stimulus which can correspond to ionic current can be related to $V(t)$ by the equation Fig. 1.27(a),

$$\frac{dV(t)}{dt} = S(t) \text{ where } S(t) > 0 \qquad ...(1.6.1)$$

6. When V reaches a critical threshold 'θ', a nerve impulse is fired.
7. After the firing, the System resets 'V' to its starting value. Suppose that a Spike is fired at time t_1 current time taken as $S(t)$ then

$$V(t) = \int_{-t_1}^{t} S(t)dt \qquad ...(1.6.2)$$

8. If the next spike is fired at t_2, then the value of θ will become

$$\theta = \int_{t_1}^{t_2} S(t)\, dt \qquad ...(1.6.3)$$

Assume that $S(t)$ is constant
Define $\delta(t) = t_2 - t_1$.
The Spike frequency is given by

$$f = \frac{1}{\delta(t)} = \frac{s}{\theta} \qquad ...(1.6.4)$$

Spike frequency is the ratio of stimulus to threshold.

This simple relation Fig. 1.27(b) looks like the one between input current and resultant firing rate in a number of sensory receptors, particularly this mold is good for limulus eye.

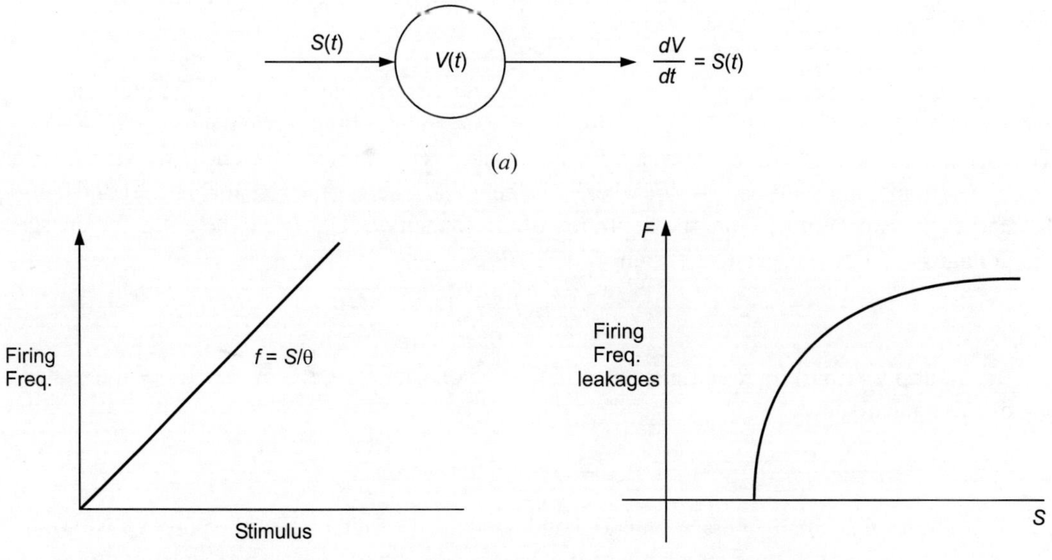

(b) *Simple Model* (c) *Forgetful Model*

Fig. 1.27

A slightly more complex version of this model considers a fall in membrane potential due to leakage. The basic equation is modified

$$\frac{dV}{dt} = -gV + S(t)$$

Avoiding complex solution, the above can be integrated and instataneous firing frequency is $\frac{\gamma}{(1-\gamma\theta/s)}$ where S is stimulus magnitude for constant input Fig. 1.27(c).

This relation (Forgetful Integrate and fire) is linear exceptat low values of S where it has threshold behaving.

A detailed analysis of this model is as follows:

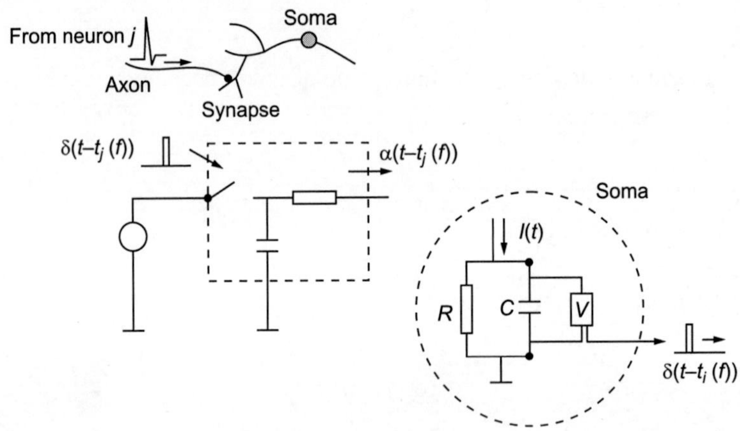

Fig. 1.28 Schematic diagram of the integrate-and-fire model. The basic circuit is the module inside the dashed circle on the right-hand side. A current $I(t)$ charges the RC circuit. The voltage $u(t)$ across the capacitance (points) is compared to a threshold ϑ. If $u(t) = \vartheta$ at time $t_i^{(f)}$ an output pulse $d(t - t_i^{(f)})$ is generated. Left part: A presynaptic spike $d(t - t_j^{(f)})$ is low-pass filtered at the synapse and generates an input current pulse $\alpha(t - t_j^{(f)})$.

The basic circuit of an integrate-and-fire model consists of a capacitor C in parallel with a resistor R driven by a current $I(t)$ Fig. 1.28. The driving current can be split into two components, $I(t) = I_R + I_C$. The first component is the resistive current I_R which passes through the linear resistor R. It can be calculated from Ohm's law as $I_R = V/R$ where u is the voltage across the resistor. The second component I_C charges the capacitor C. From the definition of the capacity as $C = q/V$ (where q is the charge and u the voltage), we find a capacitive current $I_C = C\, dV/dt$.

$$I(t) = \frac{V(t)}{R} + C\frac{dV}{dt} \qquad \text{...(1.6.6)}$$

We multiply (1.6.6) by R and introduce the time constant $\tau_m = RC$ of the 'leaky integrator'. This yields the standard form

$$\tau_m \frac{dV}{dt} = -V(t) + R\, I(t)$$

We refer to u as the membrane potential and to τ_m as the membrane time constant of the neuron.

In integrate-and-fire models the form of an action potential is not described explicitly. Spikes are formal events characterized by a 'firing time' $t^{(f)}$. The firing time $t^{(f)}$ is defined by a threshold criterion

$$t^{(f)} : V(t^{(f)}) = \vartheta$$

Immediately after $t^{(f)}$, the potential is reset to a new value $u_r < \vartheta$,

$$\lim_{t-t^{(f)},\, t>t^{(f)}} V(t) = u_r$$

For $t > t^{(f)}$ the dynamics is again given by until the next threshold crossing occurs. The combination of leakly integration (1.3) and reset (1.5) defines the basic integrate-and-fire model (Stein, 1967b). We note that, since the membrane potential is never above threshold, the threshold condition reduces to the criterion (1.4), i.e., the condition on the slope dV/dt can be dropped.

In its general version, the leaky integrate-and-fire neuron may also incorporate an absolute refractory period, in which case we proceed as follows. If V reaches the threshold at time $t = t^{(f)}$, we interrupt the dynamics (1.3) during an absolute refractory time and restart the integration at the delayed time with a new critical condition.

The idea that brain cell is computational in nature captured the attention of many researches in the period 1992-2000. According to Mc Cartly a digit physical system computes a particular function if there is appropriate mapping between the physical states of the system and the elements of the function. The stochastic behaviour of neuron is such that it maps an activation value to a binary response which is a spike in 25% of the cases and no spike in cases with probability distribution.

The minimal distance between two spikes define absolute refractory period which is followed by a phase of relative refractoriness (Gerstner and Kistfer 2002).

The model purposed by Hodking and Haxley gives a very accurate and quantitative description of the physiological data. Because of the numerous ion channels, this model is difficult for construction required in analysis. Another conceptual drawback is the complexity of number of nonlinear differential equations. It is also known that architecture of denolates play an important role in nonlinear function of the neurals.

The table shows the progress of neural models particularly in the direction of spike models.

Name of Scientists	*Origin*	*Remarks*
1. Power and Vomen	1998	Compartmental models and Genesis
2. Gerstner	1991	Model of spiking nature of neuron
3. Kistler	1997	They said about 90% of H-H Model can be stimulated by the spike model
4. Steinens and Zador	1998	Leaky integrate and fire neuron
5. Bugmann's Taylor	1997	Integrate and fire
6. Gerstner	1999	Spike response model
7. Izkevitz	2001	Resonate and fire model
8. Kistler and Gerstner	2002	Simplification of non-linear dynamic deduction of dendritic

1.6.1 Neural Communication with Spikes

(*i*) According to Adrian (1926) neurons communicate by producing sequences of fixed size electrical impulses called actions potentials or spikes.

(*ii*) According to Reike (1997) the spikes sequences are the language for which the brain is listening, the language the brain uses for its internal muscles interaction and the language it speaks as it talks to the outside world. Spikes can be coded by rate of pulses.

Rate Coding

Adrian introduced the concept of rate coding by which the number of spikes in a fixed time window exists after the application of stimulus depending on the intensity. The definition of the rate is related to sensory motor, central nervous system.

According to recent observations, the behavior of visual neurons in the cortex of the brain has temporal precisions higher than predicted by frequency coding (Abeles). This suggest that the firing rate cannot count for all the encoding of information. Recently, the focus of attention is on exploration of timing of single spikes and their relation to nervous system (Gerstner and Kistler 2002).

The application of pulse coding was suggested by Panzer 2001 by a unified approach for studying temporal, co-relation and rate coding. According to the work of Panzer and Schultz, a spike count coding exists for a narrow time window shorter than the time scale of stimulus included response fluctuations. Moreover the time windows are longer than stimulus induced characteristic time scale in the case of temporal coding phase. Various models can be classified into the see groups:

(*a*) Detailed Neural Models:

(*i*) The model of Hodgkin and Huxley is an instance of detailed model of potentials of action potential type. The complex frame works of this model which account for number of ion channels and types of synapses are difficult to construct and analyze.

(*ii*) An important drawback of these models is their complexity involving number of non-linear differentiation commutation. Simplified models are (*a*) integrated fire and (*b*) resonant and fire.

(*b*) Compartmental Models:

These are implemented by powerful simulations "Genesis" and "Neuron". For example, in each compartment a capacitance resistance circuit modelled on ordinary differentiation equations can involve 4550 components.

(*c*) Spiking Neural Models:

(*i*) The third and more important research of computation at neural science is oriented towards modelling of nature of neurons and returning essential element of the behavior in trying to simplify the complexity of descriptions.

(*ii*) The principle motivation of these models is to allow study of computational and functional principles and neural systems.

(*iii*) The reduction of detailed neural models to formal models returns to simplifications which can be summary up:

1. The non-linear model of spike generation must be reduced to a single ordinary differential Equation.
2. The special structure of the neuron (Dendritic tree) is neglected.
3. KISTLER demonstrated that spike generation in Hodgkin Huxley model can be reproduced with accuracy of 90% by a single variable model. (It was pointed out that in H-H model there is a sharp threshold like transaction between an action potential for a strong stimulus and graded response for a slightly less stimulus.)
4. Among several simplified neural models developed recently the leaky integrate and fire is the best known. It stimulates the dynamics of neuron membrane potential in response to a synaptic current by implementing an equivalent electric circuit.

5. The function of integrate and Fire circuit is to accumulate the input currents and the membrane potential reaches the max value generating a spike. (Immediately after emitting a pulse potential is reached and maintained so that there is an absolute refractory period).
6. The simplified mathematical models for spiking neurons do not account for all the computational functions of the biological neuron.
7. The integrate and fire model focuses on temporal summation.
8. The spike response model by Gerstner simplifies the action potential generation to a threshold process. The resonant and fire model focuses upon operation of neuron in a resonating mode.
9. In contrast with detailed neural models like (H-H) the computation strength of spiking neurons in these simplified models arises in the way neurons interact with each other and co-operate in large networks.
10. Other than rate coding, temporal coding can be done based on relative latencies.

1.6.2 Temporal Coding

Fig. shows coding by relative delay. This coding takes into account the differences in firing times of different neurons. This is also called delay coding or latency coding. According to flap field a vector of real numbers ($x_1 \dots x_n$) with $x_j \in (0, 1)$ can be encoded in the firing times given by t_j of n nervous in such a way that $t_j(f) = T - Cx_j$ where T reference time and Cx_j is the transmission delay.

Computational Properties of Neurons

(*i*) An essential feature of the spiking neurons is that they can out as coincidence detectors for the incoming pulses by detecting the arrival in proper time.

(*ii*) When the neuron is functioning as a coincidence detector the output firing rate is higher if the spikes arrive simultaneously as opposed to any random arrival.

(*iii*) For a neuron to work as a coincidence detector two conditions have to be satisfied.

(*a*) The post synaptic potential (the voltage response of the post synaptic neuron to a pre-synaptic action potential is referred to as post-synaptic potential) has to be evolved according to exponential decay function.

(*b*) All the transmission delay must have similar values.

Spike NNS Model

This model works on (*a*) coincidence detection and (*b*) Threshold and fire in which neurons perform linear summation of inputs similar to integrate and fire. The integration of pulses over a larger time interval is required in case of spikes effected by longer delays (1 to 10 ns for lateral synapses)

The neural model in spike NNS is a simplified version of Spike Response model called SRM suggested by Gerstor. SRM represents an alternative formulation in which instead applying differential equations to membrane potential as in Hodgkin-Huxley model a kernel based approach is adopted.

The SRM is more general than integrate and fire model because the response kernels can be arbitrarily chosen. This model produces 90% of the spike times predicted by Hodgkin-Huxley model. The dynamics of dendrite trees and nonlinear effects of synapses are taken into consideration.

In Fig. 1.29(*a*), the spikes are emitted at different moments in time defined by $t_j(f)$. The most strongly activated neuron fires first [second from left]. Its spike travels a long distance before firing of last neuron. The latenties x_j are computed with respected to T. These latencies can be computed with respect to reference time (Thorpe, 2007).

Fig. 1.29(*b*) shows the presynaptic contribution to membrane potential. Membrane voltage $V(t)$ of neuron i is considered. Each pre-synaptic neuron 'j' in 'Γ_i' emits a series of spikes Γ_j with firing time $t_j(f)$. The neuron 'i' computes the weighted sum of delayed post synaptic potential (PSP). If the sum exceeds the threshold 'θ', output spike is generated.

For Fig. 1.29(*c*) shows the membrane potential as a continuous function of time. In Fig. 1.29(*d*), the potential is taken as a discontinuous function whereas threshold θ makes the point of firing.

Governing Equation: The basic governing equation is given as:

$$V_i(t) = \eta_i(t - t_i) + \sum_{j \in \Gamma_i} \omega_{ij} \sum_{t_j(f) \in \Gamma_j} \in_{ij} (t - t_i), t - t_j^{(f)} + \int_0^{\infty} \in (t - t_i), sI^{\text{ext}}(t - s)ds$$

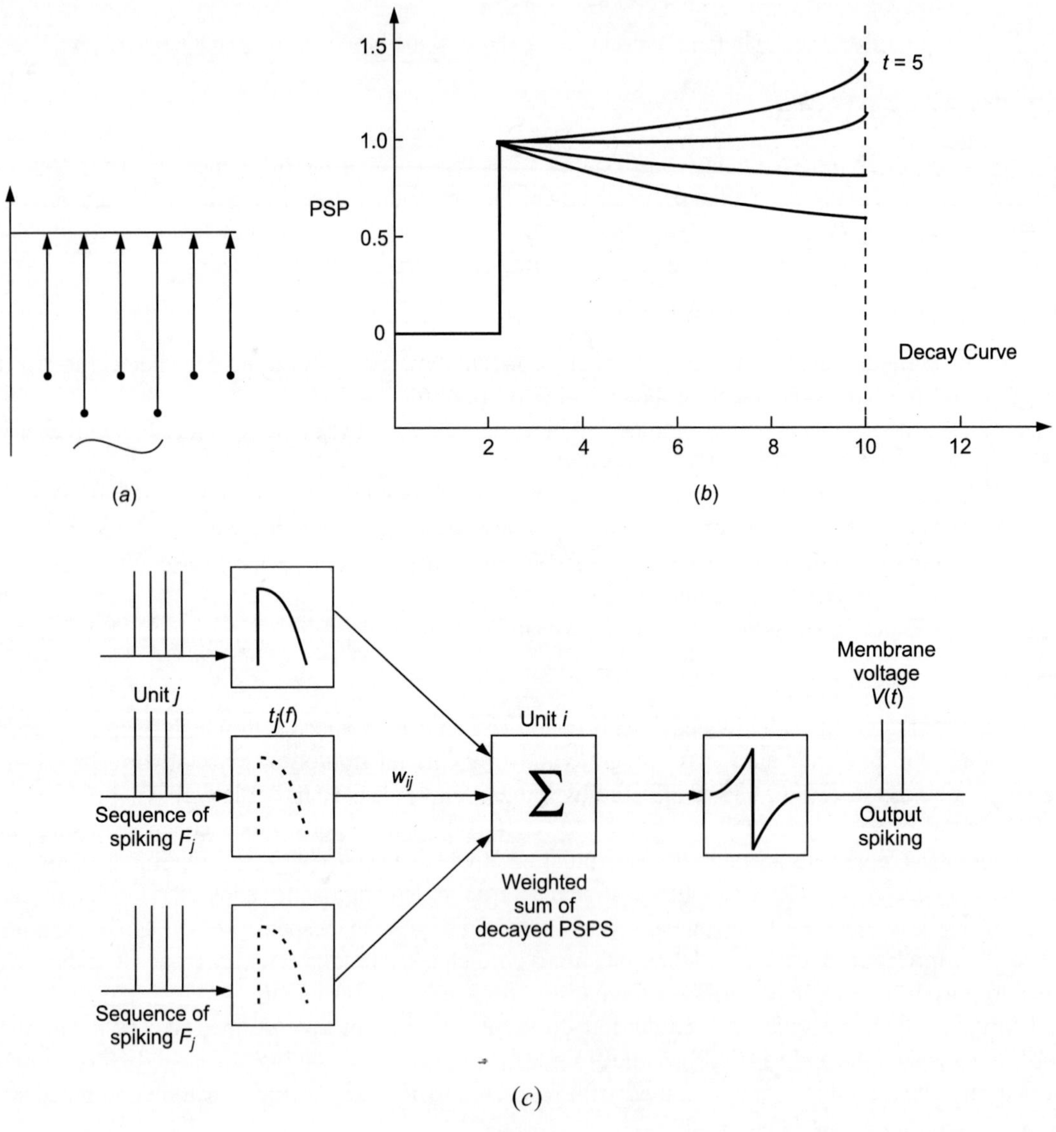

(*a*) (*b*)

(*c*)

Fig. 1.29

where

η_i = a kernel which accounts for refractoriness of the neurons

Γ_i = denotes the set of neurons pre-synaptic to 'i'

Γ_j = set of all firing neurons 'j'

ω_{ij} = synaptic strength between the cells.

The second term represents the contribution of all previous spikes occurring at $t_j(f)$ of the pre-synaptic neurons marked j on the membrane potential marked 'i'.

The kernel '$\in_{ij}$' is a function of $[t - t_j(f)]$ and this represents the time of post synaptic potential created by the pre-synaptic neuron 'j' at the time $t_j(f)$.

The evaluation of the post synaptic function depends on the time $(t - t_i)$ which has elapsed since the last spike (this is so because of the neuron is in a refractory period its response to an input spike is smaller if the neuron is fully responsive).

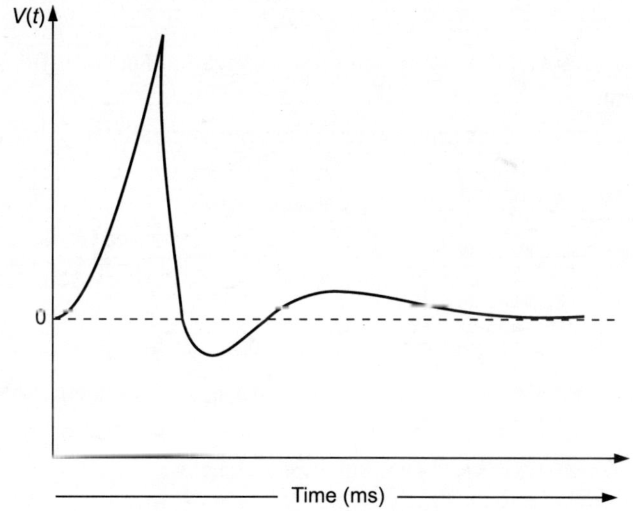

(d) Membrane Potential in SRM.

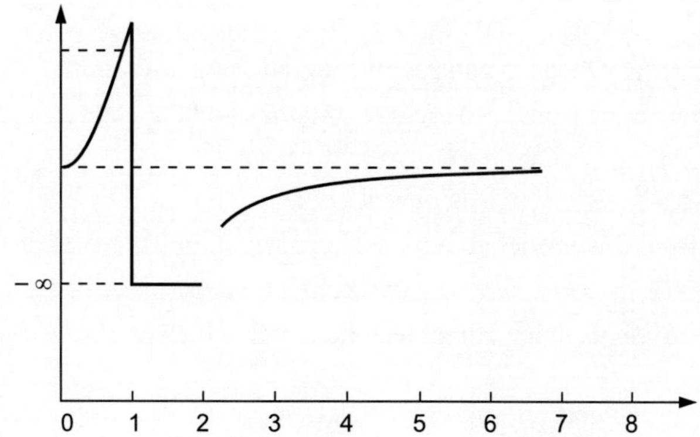

(e) Membrane Potential in Spike NNS.

Fig. 1.29

The last term represents the effect of external driving current I^{ext} and the Kernel $\bar{E}(t - t_i, s)$ is the linear response of the membrane potential to the input current which depends on the time which has passed since the last output has been sent at t_i. Before any input spike arrives, variable $V(t)$ zero. The firing of the presynaptic neuron 'j' at the time $t_j^{(f)}$ evokes a PSP given by the Kernel E_{ij}.

1.7 CHARACTERISTICS OF NEURAL NETWORKS

1. Neural networks exhibit mapping capabilities. They can map input patterns to their output patterns.
2. Neural networks learn by examples.
3. The learning procedure can be supervised, unsupervised and reinforced type.
4. Neural Networks can be trained with known examples of problems before they are tested for ability of referring.
5. Neural Networks have the capability to generalize. They can predict new outcomes from past trends.
6. They are fault tolerant and robust so that they can recall full patterns from incomplete or partial or noisy patterns.
7. Neural Networks process information at high speed in parallel and distributed manner.

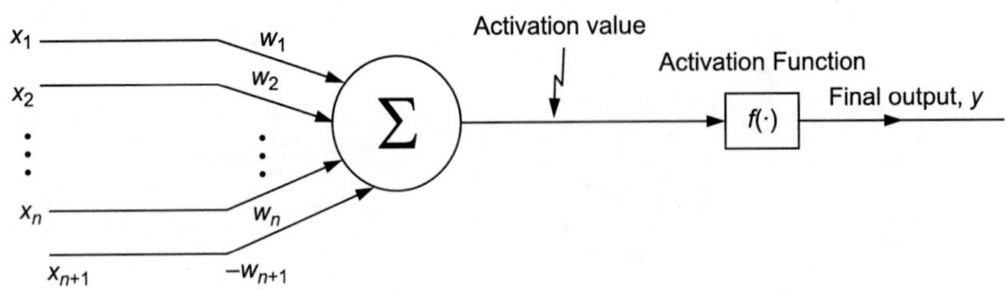

Fig. 1.30 *McCullochpits Neuron.*

1.8 McCULLOCH-PITTS MODEL

(Warren McCulloch & Walter Pitts)—1943.

(*i*) The McCulloch Pitts neuron Y can receive signal from any number of other neurons.

(*ii*) The connection weights from x_i to are x_n excitatory and the connection weights afterwards are inhibitory.

(*iii*) *Mp* neuron allows binary 0 or 1 as output.

(*iv*) In the original form this model aboves only binary output states at discrete time intervals.

(*v*) Weights are treated as constant and so there is no scope of leasing.

(*vi*) These neurons are connected by direct weighted path which can be excitatory (+ve) or inhibitory (–ve).

(*vii*) The network is associated with the threshold value.

(*viii*) The neuron fires if the net input to it is greater than threshold.

(*ix*) This threshold is set in such a way that the inhibition becomes absolute, because non-zero inhibitory input with prevent the neuron from firing.

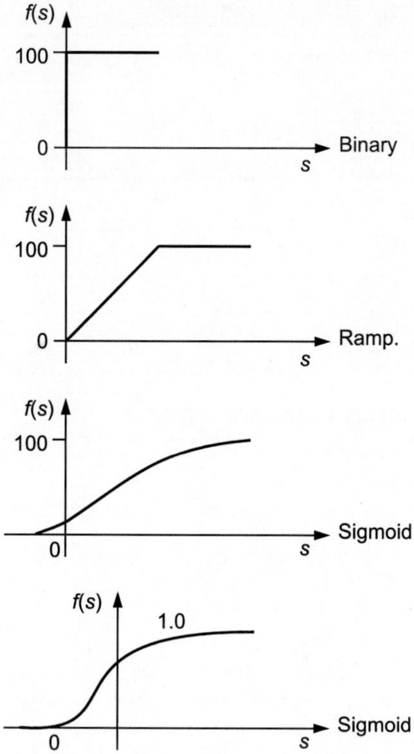

Fig. 1.31

This model is fixed on highly simplified consideration taking the biological behavior of actual neurons Fig. 1.32 shows McCulloch-Pitts model in which the input impulse at any instant. Only the binary function was used in the original model. MP neuron with single input and single output gives output after unit time. Delay property is useful in sequential logic. It can be used in memory with feedback.

1.8.1 Problems

Example 1.1: To generate Logic AND by McCulloch neuron model.

Solution

From the activation

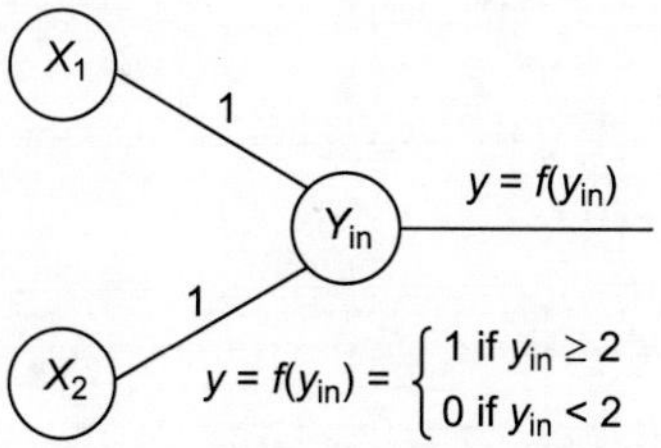

x_1	x_2	y
0	0	0
0	1	0
1	0	0
1	1	1

1. $x_1 = x_2 = 1$ $\quad y_{in} = x_1 + x_2 = 1 + 1 = 2$

$y = f(y_{in}) = 1$ since $y_{in} = 2$

2. $x_1 = 1, x_2 = 0$ $\quad y_{in} = x_1 + x_2 = 0 + 1 = 1$

$y = f(y_{in}) = 0$ since $y_{in} = 1$

3. $x_1 = x_2 = 0$ $\quad y_{in} = x_1 + x_2 = 0 + 0 = 0 < 2$

$y = f(y_{in}) = 0$ since $y_{in} = 0$

4. $x_1 = 0\ x_2 = 1$ $\quad y_{in} = 1 \quad y = 0$

Example 1.2: To generate Exclusive Logic OR gate:

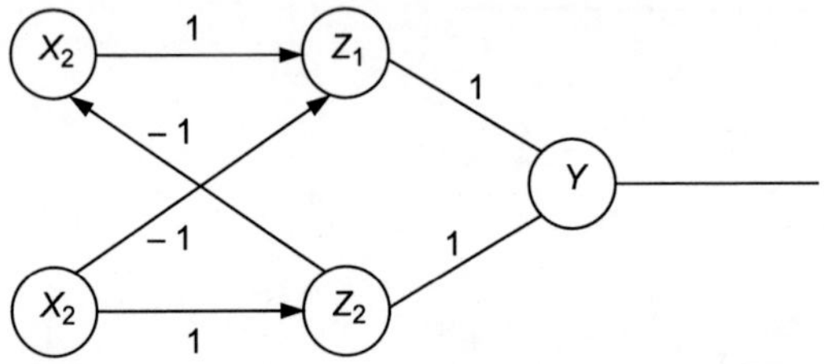

x_1	x_2	y_{in}
0	0	0
0	1	1
1	0	1
1	1	0

$$x_1\ XORx_2 = (x_1\ ANDNOTx_2)\ \text{OR}\ (x_2\ ANDNOTx_1)$$

$$x_1\ XORx_2 = z_1\ \text{OR}\ z_2$$

where

$$z_1 = x_1\ ANDNOTx_2$$

$$z_2 = x_2\ ANDNOTx_1$$

Activation

$$z_1 = (z_{\text{in}-1}) = \begin{cases} 1 & \text{if} \quad z_{in-1} \leq 1 \\ 0 & \text{if} \quad z_{in-1} < 1 \end{cases}$$

$$z_2 = (z_{in}-2) = \begin{cases} 1 & \text{if} \quad z_{in-2} \leq 1 \\ 0 & \text{if} \quad z_{in-2} < 1 \end{cases}$$

$$z_1 = x_1\ ANDNOTx_2 \quad z_{\text{in}-1} = x_1\omega_1 + x_2\omega_2$$

$$z_2 = x_2\ ANDNOTx_1 \quad z_{\text{in}-2} = x_1\omega_1 + x_2\omega_2$$

ANDNOT				*ANDNOT*			
input x_1, x_2		***output*** $= z_1$ x_1*ANDNOT*x_2		***input*** x_1, x_2		***output*** $= z_2$ x_2 *ANDNOT* x_1	
x_1	x_2	$z_{\text{in}-1}$ $\omega_1 = 1$ $\omega_2 = -1$	z_1	x_1	x_2	$z_{\text{in}-2}$ $\omega_1 = -1$ $\omega_2 = 1$	z_2
1	1	0	0	1	1	0	0
1	0	1	1	1	1	–1	0
0	1	–1	0	0	1	–1	1
0	0	0	0	0	0	0	0

Activation for output unit $y = 1$ where $Y = f(y_{\text{in}}) = \begin{cases} 1 & \text{if} \quad y_{in} \geq 1 \\ 0 & \text{if} \quad y_{in} < 1 \end{cases}$

$$y_{\text{in}} = z_1 w_1 + z_2 w_2$$

Using AND-NOT logic for x_1 and x_2 we can make the table given below:

x_1	x_2	z_1 $x_1 - x_2$	z_2 $x_2 - x_1$	y	
1	1	0	0	0	Exclusive-OR
0	1	0	1	1	
1	0	1	0	1	
0	0	0	0	0	

1.9 HISTORY

"*To make use of your minds to think conceptually is to leave the substance and attach yourself to form and discrimination between this and that a host of demons blazing forth – 9th Century Buddhist Scholor (Huang).*"

1.9.1 Early Period

1. McCulloch and Pitts (1943) suggested the unification of neuro physiology with mathematical logic. Their paper influenced von Neumann to try new designs in EDVAC (Electronic Discrete Variable Automatic Computer).
2. In 1952, Hodgkin and Huxley studied depolarization, hyper polarization "AXON" of a squid and suggested neural models to understand action potential.
3. In 1958, Rosenblatt worked on a book written by Hebb called "Organization of Behavior" which proposed a learning rule derived from a model based on synaptic connections between the nerve cells responsible for biological associative memory. An opposing paper came from Minsky and Papert in 1969 and the research work on Neural Networks came to a temporary halt.

1.9.2 Medieval Period

1. In 1977, Shaunt and Malsberg proposed self organizing map architecture.
2. In 1982, Kohonen gave self-organizing feature map networks based on localized mapping of the brain and competitive methods of learning [Hebb]. Kunihiko Fukushima in this period proposed hybrid hierarchical multilayer feed back stroke forward network employing supervised and unsupervised rules of learning.
3. In 1983, Boltzmann and Cauchy machines were proposed by Hinton. This was followed by Szu in 1987. Earlier, in 1977, James Anderson had suggested "brain state in a box" model for recurrent and auto associative network using Hebbian and gradient descent methods of learning. These machines were extension of the work of Anderson and were based on stochastic (probabilistic) networks. During this period, Hopfield Network using single layer network had come up similarly, Adaptive Resonance Theory (ART) was suggested by Carpenter and Grossberg around 1983.
4. In 1985, back propagation network was developed by David Rumelhart based on the work of Werbos (1974). Counter propagation network was given by Nielsen. Bi-directional Associative Memory (BAM) was suggested by Bart Kosko in 1988.
5. ADALINE and MADALINE were developed by Bernard Widrow in 1988. Basing on the earlier work of Widrow in 1960, MADALINE was created by a combination of ADALINE networks across multiple layers with adjustable weights.

1.9.3 Recent Developments

1. The majority of artificial neural network models were based on computational paradigm involving propagation of continuous variables from one unit to another—Rumelhart and McClelland-(1996).
2. By experimental evidence, it was suggested bioneural networks communicating through use of spikes to encode and compute information—Abeles(1991), Thorpe (1996), Reike (1997)
3. Alternative schemes consisting of rate—coding principle have been suggested to emphasize synchronization in oscillation of neurons—Reike (1991), Gerstner and Kistler (2002).
4. A new generation of pulsed neural networks had been proposed by Gerstner (1995), Stevens and Zador.
5. The neural models based on spiking nature of neurons and returning the essentials of the behavior to simplify the description were proposed by Geristner 1995 and Izhikevich in 2001.

1.10 APPLICATIONS

Artificial Neural Networks are widely used because of following reason.

(*a*) Massive parallelism.
(*b*) Distributed representation and computation.
(*c*) Learning ability.
(*d*) Generalization ability.
(*e*) Adapting.
(*f*) Inherent contextual information processing.
(g) Fault tolerance and low energy consumption.

and, so are used in following applications:

A. Pattern Recognition

Visual images, hand written characters, printed characters, speech recognition.

B. In Optimization and Constant Satisfaction

Manufacturing, scheduling and learning of travelling routes.

C. Forecasting and Risk Assessment

Meteorological prediction, weather prediction, stock market, banking sector, electrical load forecasting.

D. Control System and Robotics

Chemical and processed lauds and advanced robotics (intelligent robotics).

SUMMARY

- This unit introduces the concept of neural networks through comparison with the human system.
- Various models of neuron have been discussed. The effect of various types of stimulation has been presented in detail for the famous Hodgkin-Huxley model.
- The basics of Mc Culloch pitts model have been presented after enlistment of the characteristics.
- Few examples on MP model have been worked out.
- The history of neural networks has been given briefly, and the main applications have been listed.

Short Questions

1. Explain the terms cell body, axon, synapse and dendrite?
2. How does biological neural network work?
3. What are the main parts of the human brain?
4. What are *m*, *n*, *h* in Hodgkin-Huxley model?
5. Write the equation for integrate and fire model?
6. What is refractiveness?
7. Why EX-OR can't be executed by single neuron?
8. What are the main characteristics of AWN?
9. What is the post synaptic potential?
10. Name three applications of ANN?

Long answer Questions

1. **Explain principles of Artificial Neuron Models?**
   Refer 1.5 and 1.6 (Hodgkin-Huxley model integrate and fire neuron model)
2. **Give about characteristics of artificial neuron network?**
   Refer 1.8
3. **Explain in short, the Hodgkin-Huxley model?**
   Ref 1.5 and 1.5.1

4. **Explain briefly about spiking neuron model?**

   Ref. 1.7

5. **Explain briefly about Mc Culloch Pitts model with logic center?**

   Ref. 1.9 and 1.9.1

6. **Using Mc Culloch Pitts model implement the following with logic circuits?**

   (*i*) OR Gate

   (*ii*) AND Gate

   (*iii*) EX OR Gate

   Ref 1.91 (Problem)

7. **Explain about Organization of brain?**

   Ref.1.2

8. **Compare Human brain and Computer?**

   Ref. 1.1

9. **Write brief on historical developments of neural networks?**

   Ref. 1.10

10. **Explain about biological neural model and artificial neural model?**

    Ref.1.4

11. **Discuss applications of artificial neural networks?**

    Ref. 1.11

❑❑❑

2

ESSENTIALS OF ARTIFICIAL NEURAL NETWORKS

2.0 INTRODUCTION

This unit deals with use of artificial neurons in Neural Networks. Strictly speaking the artificial neuron has nothing to do with the biological counterpart. **It only imitates the behavior of biological neuron.** Like the dendrites of a biological neuron, input lines come to the artificial neuron with various signals of various synaptic weights. These are added and then processed linearly or non-linearly and sent to the next stages as per the requirement. By developing inter connections of such artificial neurons we can **build networks** (single layer or multilayer or recurrent). The weight adjustments can be used to make the networks learn similar to the experience gained by humans. Ultimately such adjustments or updating can be used to train the networks to achieve some results similar to the way we can train humans or animals to learn and act in some particular mode.

Various types of architecture of Neural Networks are discussed along with learning. Strategies may be **supervised or unsupervised** similar to group of students with or without a teacher. Learning rules have been developed by many researchers for various applications. Each of these rules need to be studied for a particular usage in mind. No rule is greater or inferior than the other. At the end of the unit certain applications are discussed.

This unit is preliminary study of Neural Networks. If you understand these essentials you may understand the rest without much difficulty as similar concepts are used throughout the study of Neural Networks.

2.1 ARTIFICIAL NEURON MODELS AND OPERATIONS OF ARTIFICIAL NEURON

An artificial neuron can be thought as an information processing unit which is fundamental to the operation of a neural network. The basic elements of the model of neuron is shown in Fig. 2.1.

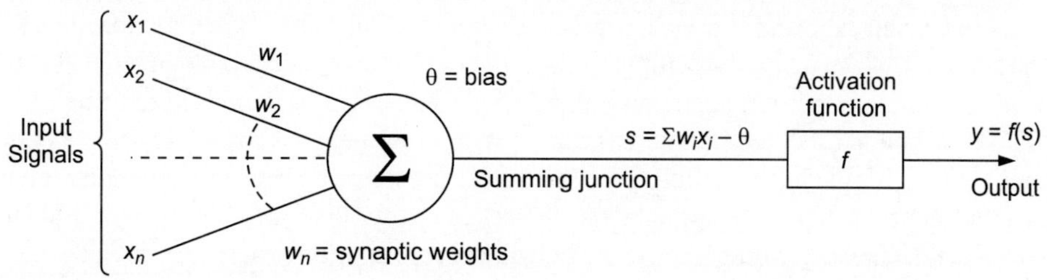

Fig. 2.1

(*a*) A set of synapses or connecting links characterized by a weight or strength. Unlike a synapse of brain, synaptic weight of artificial neuron can have negative and positive values.

(*b*) An adder (linear combiner) for the inputs.

(*c*) Activation or squashing function for limiting the amplitude.

If a bias is used then output = $\sum_{i=1}^{n} w_i/x_i + b$ and there is an Affine Transformation shown in Fig. 2.2.

Non-linear activation functions can be threshold or Heaviside type, Piecewise linear type, Sigmoid type, and effect the final output y.

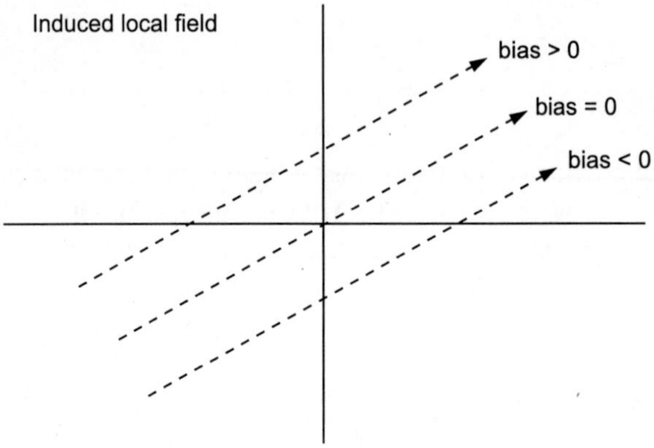

Fig. 2.2

2.1.1 Stochastic Model

The above model is deterministic because the input-output relation is precisely defined for all the input. If $p(s)$ is the probability of neuron to be in +1 state and $1 - p(s)$ is the probability of neuron to be in –1, (1, –1 are assumed to the possible states). This $p(s)$ is often chosen to be $\frac{1}{1+e^{-(s/T)}}$ where T is called pseudo temperature. T is a parameter which controls the thermal fluctuations representing the effects of synaptic noise. When $T \to 0$, the stochastic neuron reduces to a noiseless or deterministic model.

2.2 OPERATIONS OF ARTIFICIAL NEURON

The combination of all $w_i x_i$ and the bias is processed through an activation function unit to give the output. The function of the activation unit is to limit the output in an acceptable manner for the logic to be used. Under suitable condition decided by the design, the neuron is supposed to "fire".

The output of one neuron can be interconnected with other neurons to build networks required for various applications. Single layer or multi layer neuron structures are prevalent depending on the usage. One set of input neurons can be separated from the set of output layer of neurons by a hidden layer of neurons. Artificial Neural Networks (ANN) are non-linear information processing devices built from the interconnected elementary processing devices (neurons).

The basic technique of weight adjustment gives lot of freedom to the designer to build ANN. In many practical applications initial choice of the weights could be random. By feedback or another method, these weights are changed so as to achieve desirable goals by a process of iteration.

The weight adjustment is comparable to the process of human learning by experience. Analogously, the process of weight variation is termed as 'learning' of the Artificial Neural Networks. Conversely, we may say that we can train the Artificial Neural Networks to achieve a certain goal as a human can be trained to attain higher objectives.

2.3 TYPES OF NEURON ACTIVATION FUNCTION

Similar to the biological neuron an artificial neuron receives input representing the output of other neurons. The inputs are multiplied by the weights analogous to the synaptic strength. All these weighted inputs are added (s) and made to pass through activation function which determines the output. These can be linear or non-linear. The commonly used Activation Functions are listed below:

1. HEAVISIDE (THRESHOLD)

$$f(s) = 1 \quad \text{if} \quad s \geq \theta$$
$$= 0 \quad \text{if} \quad s < \theta$$

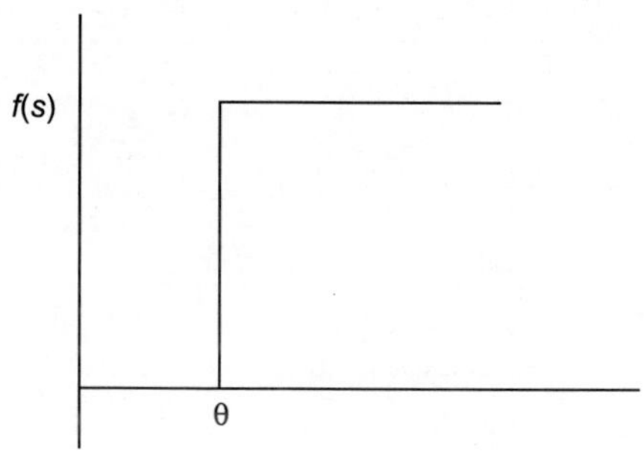

(*a*)

2. LINEAR

$$f(s) = s$$

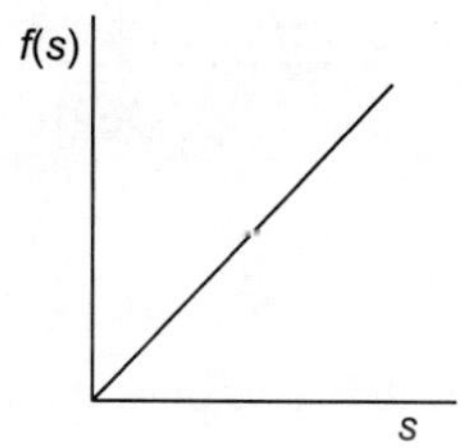

(*b*)

3. PEICEWISE LINEAR

$$f(s) = 1 \quad \text{if } s > \theta$$
$$= s \quad \text{if} - \theta < s < \theta$$
$$= -1 \quad \text{if } s < -\theta$$

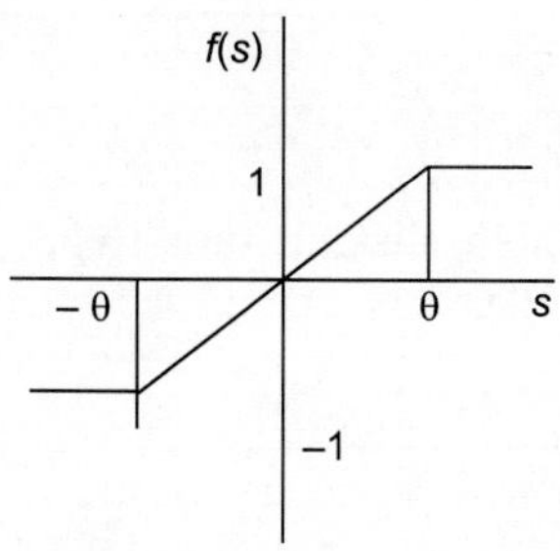

(*c*)

4. HARD LIMITER (SIGNUM)

$$f(s) = sgn\ (s)$$
$$= +\ 1 s > 0$$
$$= -\ 1 s > 0$$

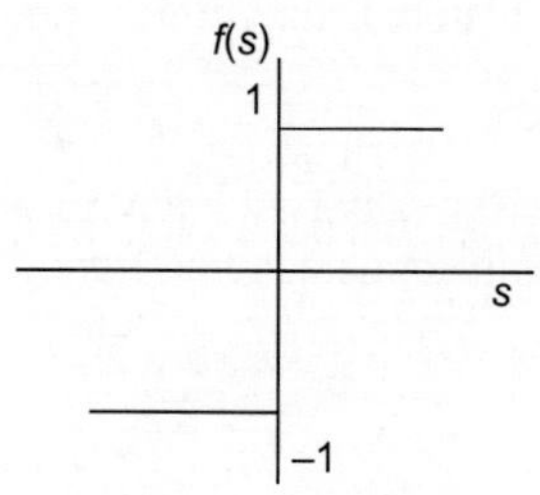

(*d*)

5. UNIPOLAR SIGMOIDAL

$$f(s) = \frac{1}{1+\exp^{-\lambda s}}$$

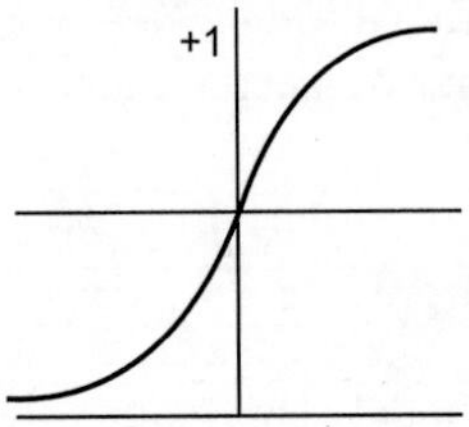

(e)

6. BIPOLAR SIGMOIDAL (Hyperbolic Tangent)

$$f(s) = \frac{2}{1+\exp^{-\lambda s}} - 1$$

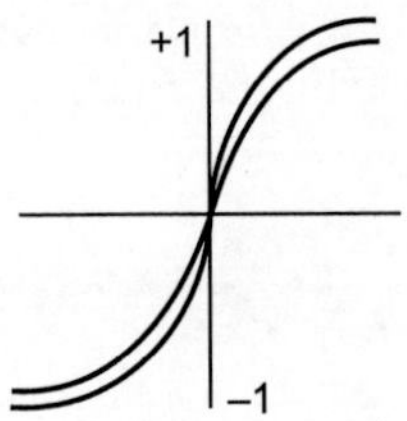

(f)

Fig. 2.3

7. UNIPOLAR MULTIMODAL

$$f(s) = \frac{1}{2}\left[1+\frac{1}{M}\sum_{m=1}^{M}\tanh(g^{m}(1-w_0^{m}))\right]$$

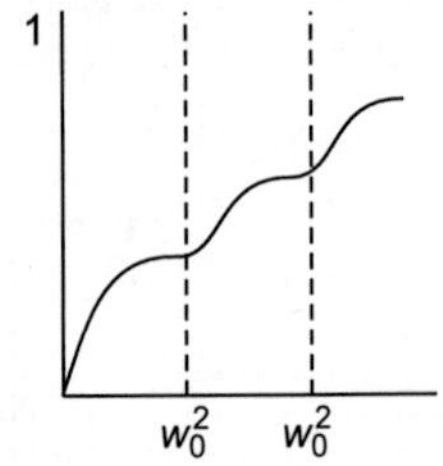

8. RADIAL BESSEL'S FUNCTION

$$\text{Output} = \exp(I)$$

where $$I = \left[\frac{-\Sigma(W_i(t) - X_i(t)^2}{2\sigma^2}\right]$$

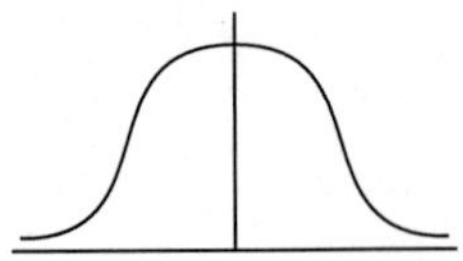

It may be noted that sigmoid functions are continuous and differentiable. Bipolar Sigmoidal is equivalent to Tanh function. In comparison with Sigmoidal function Hard Limiter functions are useful for discrete application. Activation functions are also referred to as signal function in particular applications.

Some Signal Functions

1. Logistic signal: $s(x) = \dfrac{1}{1+e^{-cx}}$

   (Gibbs probability density function) It has maximum statistical entropy.

2. Threshold signal function: $s(x^{k+1}) = \begin{cases} 1 \text{ if} & x^{k+1} > T \\ s(x^k) & x^{k+1} = T \\ 0 & x^{k+1} < T \end{cases}$

   For an arbitrarily real value threshold T, where k denotes discrete time step.

3. Hyperbolic tangent signal function:

$$s(x) = \tanh(cx)$$

$$s' = c(1 - s^2) > 0$$

$$\tanh(cx) = \frac{e^{cx} - e^{-cx}}{e^{cx} + e^{-cx}}$$

4. Threshold linear signal function:

$$s(x) = \begin{Bmatrix} 1 & \text{if } cx \geq 1 \\ 0 & cx < 0 \end{Bmatrix}$$

$$s = (x = \min (1, \max (0, cx))$$

$$s' = c > 0$$

5. Exponential distribution signal function:

$$s(x) = \max (0, 1 - e^{-cx})$$

$$s' = ce^{-cx} > 0$$

6. Ratio polynomial:

$$s(x) = \max\left(0, \frac{x^n}{c + x^n}\right)$$

$$s' = \frac{c_n x^{n-1}}{(c + x^n)^2} > 0$$

2.4 ARTIFICIAL NEURAL NETWORKS ARCHITECTURE

A simple method of representation Artificial Neural Networks architecture is shown as Fig. 2.4.

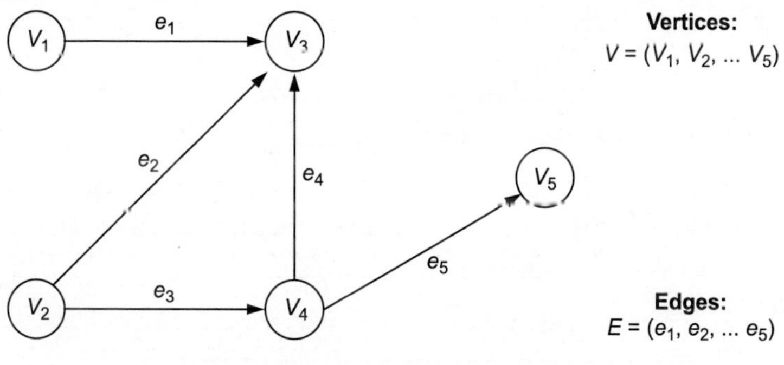

Fig. 2.4 *Directed Graph (2-Tuple)$_z$*

1. Artificial Neural Networks are defined as a data processing system consisting of a large number of simple, highly interconnected processing elements (Artificial Neurons) in an architecture inspired by the structure of Cerebral Cortex of the brain.
2. Generally an Artificial Neural Networks structure can be represented by a directed Graph (shown in Fig. 2.4).
3. A single directed graph (2-Tuple) consists of a set of vertices and a set of edges.
4. When each edge is assigned an orientation, the graph is directed and is called directed graph or dia-graph.
5. These are called Diagraphs because signals in Artificial Neural Networks are restricted to flow in specific direction.
6. The vertices of the graph may represent neurons (input or output) and edges may represent the synaptic links. The edges are labelled by the weights attached to synaptic lines.

2.5 CLASSIFICATION OR TAXONOMY OF ANN-CONNECTIVITY

1. According to the learning mechanism there are three basic classes of networks.
   (*a*) Single Layer Feed Forward network has two layers—input layer and output layer. The synaptic linkages connect every input neuron to the output neuron (and not vice versa). This network is called feed forward or Acyclic. In spite of having two layers, this is termed single layer because the output layer take part in computation (Fig. 2.5(*a*)).
   (*b*) In Multilayer Feed forward layer there is an intermediate layers called hidden layer which aids in performing some important computation before these are passed on to the output layers. There are two types of weights—input to hidden layer and hidden layer to output.
   Multilayer feed forward network (Fig. 2.5(*b*)) link output neurons m_1 neurons in the first hidden layer and m_2 neurons in the next hidden layer and n output neurons is termed $1 - m_1 - m_2 - n$.
   (*c*) The recurrent networks have at least one feedback loop there can also be neuron with self feed.

   Some important architectural behaviors of neural networks are
   (*a*) Mapping capability. (*b*) Learning by example.
   (*c*) Capacity to generalize. (*d*) Robust and fault tolerant design.
   (*e*) Parallel and distributed operation.

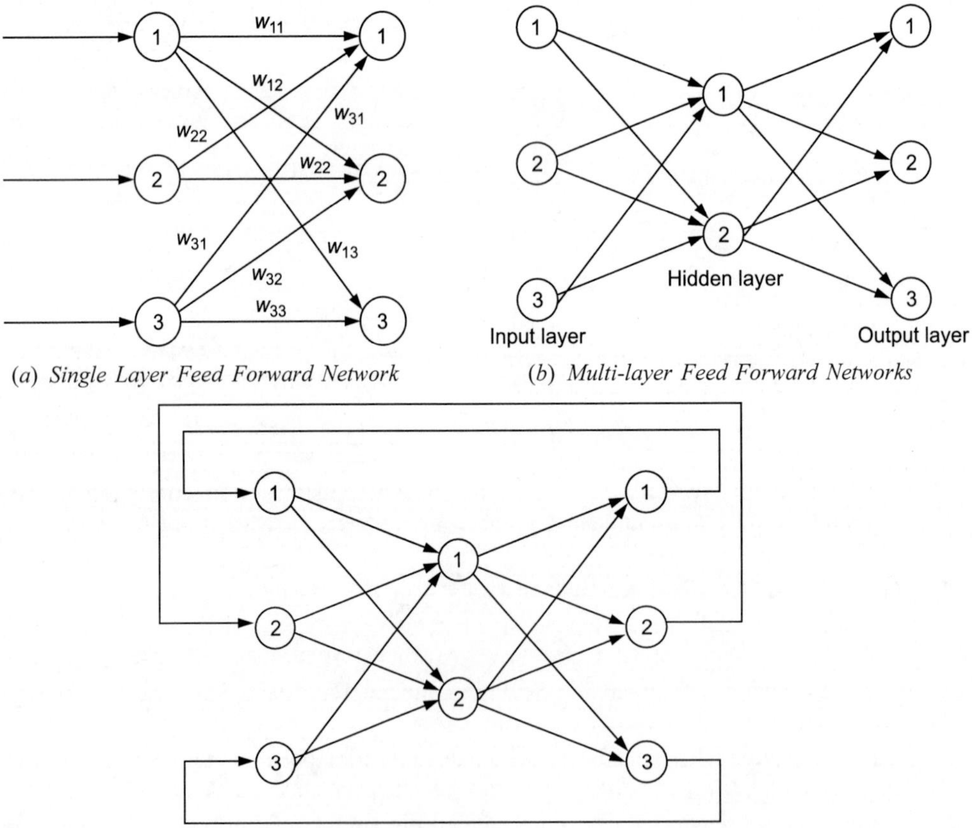

(*a*) Single Layer Feed Forward Network (*b*) Multi-layer Feed Forward Networks

(*c*) Recurrent Neural Network

Fig. 2.5

Classification of learning strategies:

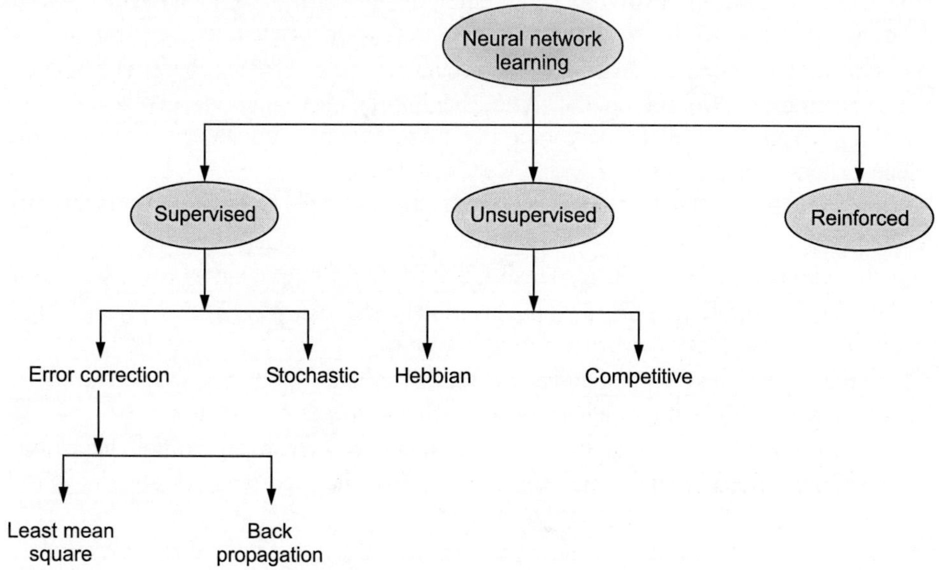

Table Showing Classification According to Learning Techniques and Architecture

	SLLF	*MLFF*	*RECURRENT*
GRADIENT DESCENT	ADALINE Hopfield perception	CCN MMLF RBF	RNN
HEBBIAN	AM Hopfield	Neocognition	BAM Hopfield BSB
COMPETITIVE	LVQ SOFM	—	ART
STOCHASTIC		—	BOLTZMANN CAUCHY

TAXONOMY of ANNS CONNECTIVITY

ADALINE = ADAPTIVE LINEAR NEURAL ELEMENT
ART = ADAPTIVE RESONANCE THEORY
AM = ASSOCIATIVE MEMORY
BAM = BIDIRECTIONAL ASSOCIATIVE MEMORY
BSB = BRAIN STATIC IN A BOX
CCN = CASCADE CORRELATION
LVQ = LINEAR VECTOR QUANTISATION
MADALINE = MANY ADALINE
RBF = RADIAL BASIC FUNCTION
RNN = RECURRENT NEURAL NETWORKS
SOFM = SELF ORGANISED FEATURE MAD

2.6 NEURAL DYNAMICS

1. The activation function involves small membrane pulses. Signals involve large Axonal pulses. Both of these involve electrical pulses of potential difference and temporal summation.
2. The potential difference in electric charge occurs between the inside charge and outside neurons surface membrane. The neuron transmitters having chemical signals from other neurons through abutting synapses modulate membrane potential. Synapses transducer, converts pulse coded electrical neuronal signals into neuron transmitter signals.
3. The membrane potential differences/pulses accumulate at the Axon hillock on where the neuron is connected to the Axon or its long branches.
4. It is the Axon hillock which is responsible for firing the output signal.
5. If the potential difference pulse arriving at Axon hillock exceeds a Time Varying Threshold, the Axon hillock membrane generates a signal pulse or AP.
6. The large signal pulse propagates down the Axon and its braches where Axonal insulators restore and amplify as it propagates till it reaches the synaptic junction.
7. Synaptic transducer processes the electrical signal into chemical emitter (neuro-transmitter) signal which affects the potential difference across the membrane of abutting post-synaptic neuron.
8. At molecular level, ions drive charges making changes in potential differences.
9. Ionic flow changes the Axon hillock membrane potential by modifying the membrane conductance. Conductance change involves selectively, opening and closing the molecular channels. Na^+ and K^+ ions flow across the membrane through the channels.
10. Deficiency of +ve ions in the protoplasm indicates signals from synaptic connections. Additional signals result in temporary depolarization of the resting potential.
11. When the potential which is a level of above –60 V to 65 V, the membrane suddenly loses its impermeability and Na^+ enter into the protoplasm.
12. This sudden change in the membrane potential causes the neuron to fire. The membrane slowly recovers its original properties and regenerates the resting potential in a period of few milli seconds.
13. During this recovery period, the neuron remains fully or partly in capable of further excitation.
14. The discharge which initially occurs in the cell body propagates an along the Axon to the synopsis.
15. The intensity of the signal is encoded in the frequency of sequence of pulses of activity, which range from 1 to 100/sec. [up to 300 sec.]
16. The speed of propagation of signals in discharge is in the range (0.5 – 2 m/sec).
17. This signal stops at the synapses as there is no direct conducting link to the next neuron. Transmission of signal across synaptic gap is made effective by chemical activity. When the signal arrives at pre synaptic nerve terminal (Synaptic nerve terminal button) special substances called neuro transmitters are produced in tiny amounts.
18. The neuro-transmitter molecules travel across the synaptic junction reaching the post-synaptic neuron within 0.5 m/sec. These neuro-transmitters modify the conductance of the post-synaptic membranes for certain ions, causing depolarization of the post, pre-synaptic potential.
19. If the induced polarization potential is positive, the synapse becomes excitatory and if it is negative synapse becomes inhibitory.
20. Na^+ and K^+ change the conductance and ability to follow across the membrane by altering molecular channels in a complex way. Na^+ and K^+ are antagonistic to each other.

21. Concentrations of Na^+ and K^+ have opposite effects on polarization status. Sodium pump is used for maintaining proper Na^+ content. Let of energy (metabolic) is expended in such costly exercise.
22. The complexity of the human central nervous system is due to vast number of neurons and their interconnection.
23. The connectivity is characterized by complementary properties of convergence and divergence.
24. The human cortex has a converging input of the order of 10^4 synapses. The total number of neurons in the cortex is distributed in the layers of vertical tissue with a density of 15×10^4 neurons/mm^2.
25. There are about 10^{15} synaptic connections in the human brain, the majority of which have been developed in the first few months of birth.
26. Activation functions are normally made non-linear to make it similar with non-linear behavior of conduction current in biological neuron. However, as artificial neuron is not to be taken as a exact copy of a biological neuron, many forms of non-linear function and Sigmoidal function are very similar to the function of biological neuron.
27. The term signal velocity refers to $\frac{ds}{dt}$ in which s stands for various signal functions.
28. This signal function are induced by activation and are dependent on firing frequency of AP or the number of pulses in a sampling interval.

2.7 LEARNING STRATEGY

2.7.1 Supervised Learning

The method of setting the values of weight is called the process of learning or training. The techniques of modifying the weight in the connection between the layers of the network with the objective of achieving an expected output constitute the training procedures. Learning strategy can be supervised, unsupervised and reinforcement type.

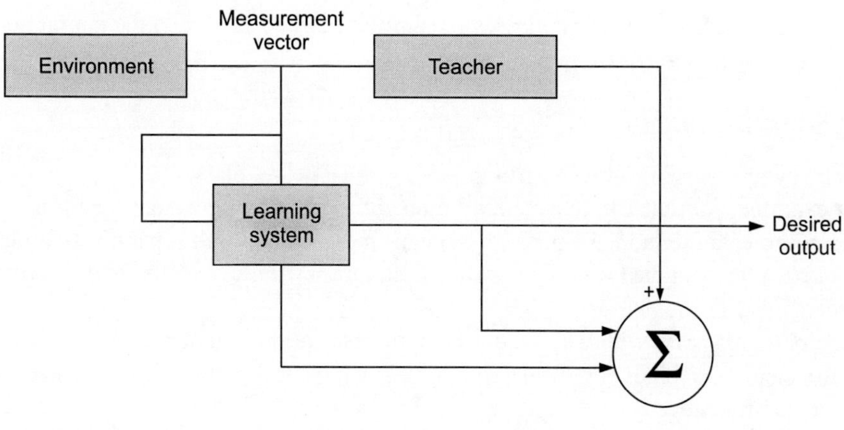

Fig. 2.5 *Supervised learning*

Unsupervised Learning

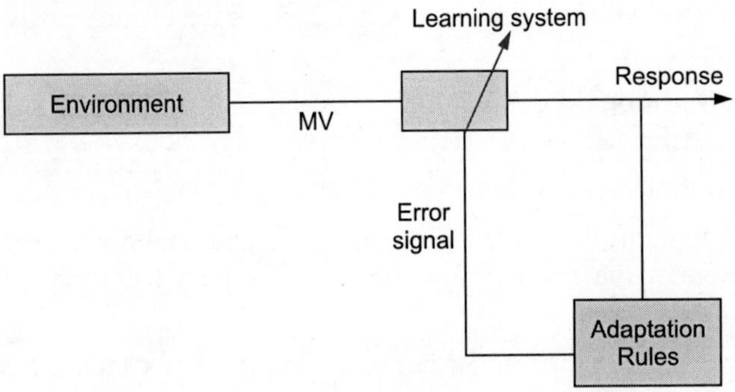

Fig. 2.6

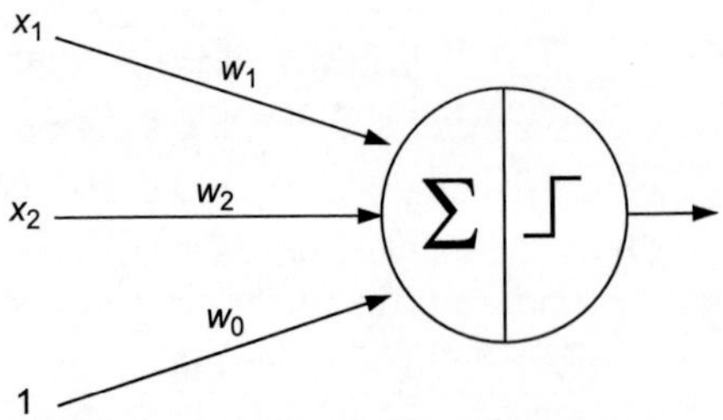

Fig. 2.7

In the above Fig. 2.7.

$$\text{Net stimulus } (N_s) = w_0 + w_1 x_1 + w_2 x_2$$

$$\Rightarrow \quad x_2 = \frac{w_1 x_1}{w_2} + \frac{w_0 - N_s}{w_2}$$

$$w_0 \rightarrow \text{bias factor}$$

The above equation represents a straight line which can be shown as:

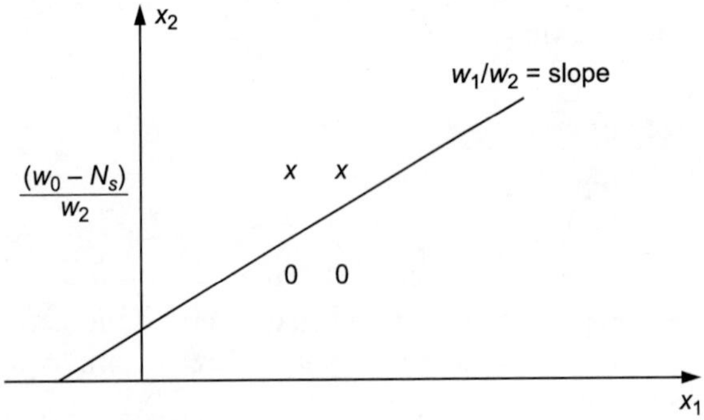

Fig. 2.8

This line can be used to divide two classes of samples marked (x, 0) and the neural network can be thought as a basic linear classifier.

2.7.2 Decision Boundary

1. Each shape represents objects, which belong to same set class.
   For example, A circle can represent all men and ellipse can represent the birds.
2. The neuron (in the mode of perceptron) can find a line, which separates two classes.
3. The lines represent the locus of the neuron activation values possible for a given set of weights associated with the neuron.
4. For a particular case with weighted inputs, the activation function must lie on the line.
5. The decision machines (or) decision engines, which follow this type of equation, are called linear classification.
6. By changing w_0, w_1, w_2 and the net stimulus value, the position of the line can be moved until one side of line contains examples of a given class and the other side contains the other class. The process of change w_0, w_1, w_2 is a part of training program.

7. When more than two dimensions are involved, the equation becomes a representation of a hyper plane of separation.
8. The data involved in the above process are called linearly separable data as compared to non-linearly separable data.
9. A perceptron has a special algorithm for linear by separate. It adjusts the weights using the error vector between data point and a separating decision line.

2.7.3 Decision Line

Change in weights are made as per the training program until none of the trained data produces error. The Perceptron learning rule requires that a desired output is known before the network can adopt the weights to find the line which separates one class from the other.

2.7.4 Supervised Learning

This process of using desired outputs for training the neural network is known as supervised learning.

1. The supervised learning employs a teacher to assist the network by telling the network what the desired response to a given stimulus is to be.
2. The operation is similar to a student being guided by a teacher. Ref to Fig 2.5. Learning system is exposed to the environment which is represented by measurement vector (MV) of features. MV is also presented to a teacher who by experience determines the desired response.
3. This desired response is used to create error signal which adapts weights of the learning system.
4. Each input vector has an associated desired output vector which is used for the training.
5. In case of supervised learning every input vector has a corresponding desired output.
6. This is not practicable in all cases.

Applications

1. Supervised learning is used for pattern association required in pattern recognition. If a neural network is trained to associate a set of input vectors with a corresponding set of output vectors then it is called ASSOCIATIVE MEMORY NET.
2. If output is same from input, then the network is called AUTO ASSOCIATIVE MEMORY. If output is different from input then the network is called HETERO ASSOCIATIVE MEMORY.
3. Some of the supervised learning algorithm are HEBB net, pattern association memory net, back propagation net, counter propagation net.

2.7.5 Unsupervised Learning

1. The main difference is that no teacher is employed.
2. It is similar to students learning on their own.
3. It is somewhat open loop with a set of adaptation rules.
4. The MV is fed to Learning System and the system response is obtained.
5. Based on the system response and adaptation rule, the weights of the Learning System are adjusted by the error signal to obtain the desired performance.
6. The adaptation rule generates the error signal which is done by the teacher in supervised learning system.

Applications

(*a*) Self Organizing Map (SOM).
(*b*) Adaptive Resonance Theory (ART).

2.7.6 Reinforcement Training

1. In this method, a teacher is assumed to be present but the right answer is not given to the network.
2. Instead, the network is given an indication of whether output answer is right or wrong.
3. The networks use this information (indication) to improve its performances.
4. If sufficient information is available, reinforced training can handle a many specific problems.

Comparison of Supervised Learning and Un-Supervised Learning

Supervised Learning	*Unsupervised Learning*
1. Learning associated with a teacher. 2. Resulting error can be used to correct parameters for improving network. 3. Learning algorithm is accurate. 4. This refer to "search" in weight space along error gradients. 5. There is user class membership concept.	1. There is no teacher 2. Rules of adaptation are used for improvement. 3. Learning algorithm is comparatively less accurate. 4. This refer modifications of parameters is quasi biological method. 5. There is no class membership basis.

Batch Learning Vs Stepwise (Incremental) Learning: Batch Learning refers to the case in which the weight is adjusted in a signal step using complete set of input, output data. This does not use feedback information. In case of incremental learning feedback from teacher on environment is used stepwise method. It may be noted that only the fan-in weight vector of the winning neuron is adjusted. Winning neighbourhood can be extended beyond single neuron also. Initialization can be random values with normalized lengths.

2.8 LEARNING RULES

A neuron can be considered to be an adaptive element in the sense that its weights can be modified depending on the input signal, output value and the response of the teacher. In case of unsupervised learning (without teacher) the modifications will be based on input and/or output.

Let w_{ij} be the weight corresponding to the input and i^{th} neuron as a part of the weight vector w_i. The jth input can be from another neuron or external source.

The activation functions can be different as per the learning rules. A threshold parameter can be introduced as one of the weights (for example x_n may be assigned a value of –1)

In general, the weight vector $\boldsymbol{w}_i = [w_{i1}\ w_{i2} \ldots w_{in}]^T$ increases in proportion the product of input x and learning signal r which is a function of $\boldsymbol{w}_i$, $\boldsymbol{x}$ and occasionally teacher is input d_i.

Assuming a positive number a learning constant c we can write the general learning rule as

$$\Delta\, \boldsymbol{w}_i(t) = cr\ [\boldsymbol{w}_i(t), \boldsymbol{x}(t), d_i(t)]\ \boldsymbol{x}(t) \qquad ...(2.8.1)$$

The weight vector adapted at time t changes at the next instant (learning step) by

$$\boldsymbol{w}_i\,(t+1) = \boldsymbol{w}_i(t) + \Delta\, \boldsymbol{w}_i(t) \qquad ...(2.8.2)$$

For the kth step (in discrete case) the equation can be rewritten as

$$\boldsymbol{w}_i^{k+1} = \boldsymbol{w}_i^k + cr\,(\boldsymbol{w}_i^k, \boldsymbol{x}^k, d_i^k)\,\boldsymbol{x}^k \quad ...(2.8.3)$$

For continuous time learning the change can be expressed as $\dfrac{d\boldsymbol{w}_i}{dt} = cr\;x(t)$

2.8.1 Hebbian Learning Rule

This represents a case of feed forward, unsupervised learning. In 1949 Hebb postulated that

"When an axon of cell A is near enough to excite a cell B and repeatedly or persistently takes part in firing it, some growth process or metabolic change takes place in one or both cells such that A's efficiency, as one of the cells firing B, is increased."

This rule basically implies that if the cross product of output and input is positive there is an increase in weight w_{ij}. In case the cross product or correlation term is negative, the weight decreases. Frequent input patterns can eventually produce the largest output.

The learning signal is a function of $\boldsymbol{w}_i^T x$. So increment $\Delta \boldsymbol{w}_i$ becomes $cf(\boldsymbol{w}_i^T x)\boldsymbol{x}$. Single weight W change can be computed by

$$\Delta\, w_{ij} = cf(\boldsymbol{w}_i^T x)x_j = cf(s_i)x_j \text{ (assuming zero threshold)} \quad ...(2.8.4)$$

A basic assumption for this rule is that the weights may be initialized around $w_i = 0$

EXAMPLE 2.1:

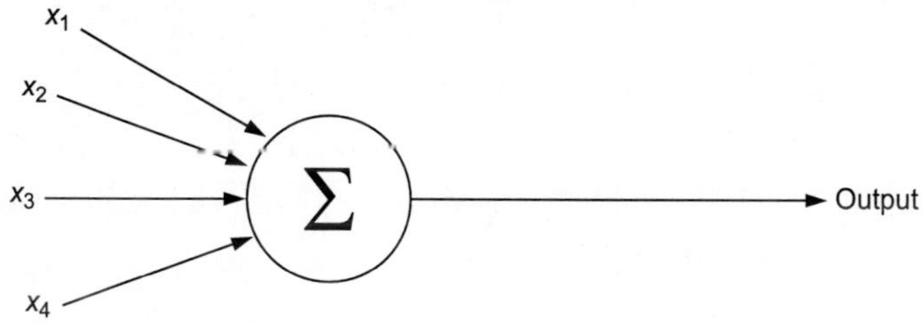

For the neuron shown assume an initial weight vector $\boldsymbol{w}^1 = \begin{bmatrix} 1 \\ -1 \\ 0 \\ 0.5 \end{bmatrix}$

the training is to be done with $\boldsymbol{x}_1 = \begin{bmatrix} 1 \\ -3 \\ 1.5 \\ 0 \end{bmatrix}$ and $\begin{bmatrix} 1 \\ -0.5 \\ -3 \\ -1.5 \end{bmatrix} = \boldsymbol{x}_2$

The Learning constant is 1. Bipolar Sigmoid function considered.

Step 1: Input x_1 is applied:

$$\boldsymbol{w}^{1T}\boldsymbol{x}_1 = [1 \;\; -1 \;\; 0 \;\; 0.5]\begin{bmatrix} 1 \\ -3 \\ 1.5 \\ 0 \end{bmatrix}$$

$$= 1 \times 1 + (-1) \times (-3) + 0 \times 1.5 + 0.5 \times 0 = 4$$

$$\text{Sgn}(\boldsymbol{w}^{1T}\, x_1) = 1$$

Updated weights are $\boldsymbol{w}^2 = \boldsymbol{w}' + \text{Sgn}(\boldsymbol{w}'^T \boldsymbol{x}_1)\, \boldsymbol{x}_1 = \boldsymbol{w}' + \boldsymbol{x}_1$

$$= \begin{bmatrix} 1 \\ -1 \\ 0 \\ 0.5 \end{bmatrix} + \begin{bmatrix} 1 \\ -3 \\ 1.5 \\ 0 \end{bmatrix} = \begin{bmatrix} 2 \\ -4 \\ 1.5 \\ 0.5 \end{bmatrix}$$

Step 2: x_2 is the input:

$$w^{2T}\, \boldsymbol{x}_2 = [2 \;\; -4 \;\; 1.5 \;\; 0.5] \begin{bmatrix} 1 \\ -0.5 \\ -3 \\ -1.5 \end{bmatrix}$$

$$= 2 \times 1 + (-4) \times 0.5 + 1.5 \times (-3) + 0.5 \times (-1.5) = -1.25$$

$$\text{Sgn}(\boldsymbol{w}^{2T}\, \boldsymbol{x}_2) = -1.75$$

So $$\boldsymbol{w}^3 = \boldsymbol{w}^2 + \text{Sgn}(\boldsymbol{w}^{2T}\, \boldsymbol{x}_1)\, \boldsymbol{x}_1 = w^2 - x_2$$

$$= \begin{bmatrix} 2 \\ -4 \\ 1.5 \\ 0.5 \end{bmatrix} - \begin{bmatrix} 1 \\ -0.5 \\ -3 \\ -1.5 \end{bmatrix} = \begin{bmatrix} 1 \\ -3.5 \\ 4.5 \\ 2.0 \end{bmatrix}$$

If a continuous bipolar activation is to be used then $f\,(\Sigma\; w'^T \boldsymbol{x}_1)$ will be considered in Step 1.

2.8.2 Perception Learning Rule

This is a case of supervised learning. According to Rosenblatt (1958) the learning signal r is the difference between the desired response (d_i) and actual response outputs which is $\text{sgn}(s_i)$ or $\text{sgn}(\boldsymbol{w}_i{}^T x)$. The weight changes are done by

$$\Delta\, \boldsymbol{w}_i = c[d_i - \text{sgn}(\boldsymbol{w}_i{}^T \boldsymbol{x})]\boldsymbol{x} = c[d_i - \text{sgn}(s_i)]\boldsymbol{x} \qquad ...(2.8.5)$$

or $$\Delta\, w_{ij} = c[d_i - \text{sgn}(\boldsymbol{w}_i{}^T \boldsymbol{x})]x_j \quad \text{for } j = 1, 2, \ldots n \qquad ...(2.8.6)$$

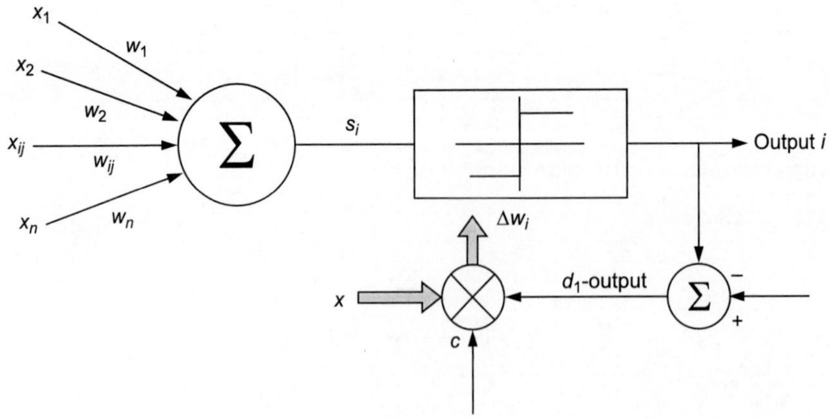

Fig. 2.9

The weights are adjusted if output is incorrect. As the desired response is either +1 or –1. $\Delta w_i = \pm 2cx$. Positive sign applies when $d_i = 1$ and $\text{sgn}(s_i) = -1$ .The Negative sign applies when $d_i = -1$ and $\text{sgn}(s_i) = -1$. There is no adjustment when d_i output. In this case, the weights can be of random initialization.

EXAMPLE 2.2: The set of input training vectors are $x_1 = \begin{bmatrix} 1 \\ -2 \\ 0 \\ -1 \end{bmatrix}, x_2 = \begin{bmatrix} 0 \\ 1.5 \\ -0.5 \\ -2 \end{bmatrix} \quad w^1 = \begin{bmatrix} 1 \\ -1 \\ 0 \\ 0.5 \end{bmatrix}$

$d_1 = -1, d_2 = -1 \quad c = 0.1$

Step 1: Input is x_1, output desired is –1

$$w^{1T} x_1 = [1, -1, 0, 0.5]\begin{bmatrix} 1 \\ -2 \\ 0 \\ -1 \end{bmatrix} = 2.5$$

Sgn(2.5) = 1. So correction is necessary as d_1 is not the same as outputs

$$w^2 = w^1 + 0.1\ (-1\ -1)\ x_1 = \begin{bmatrix} 1 \\ -1 \\ 0 \\ 0.5 \end{bmatrix} - 0.2\begin{bmatrix} 1 \\ -2 \\ 0 \\ -1 \end{bmatrix} = \begin{bmatrix} 0.8 \\ -0.6 \\ 0 \\ 0.7 \end{bmatrix}$$

Step 2: Input is x_2 desired output = –1

$$\text{Sgn}(w^{2T} x_2) = \text{sgn}(-2.3) = -1$$

Correction is not necessary as the outputs are matching ($d_2 = \text{sgn}(-2.3)$).

2.8.3 Delta Learning Rule

This is applicable for continuous activation functions and in supervised training. The learning signal r called delta is given by $[d_i - f(w_i^T x)]\, f'(w_i^T x)$.

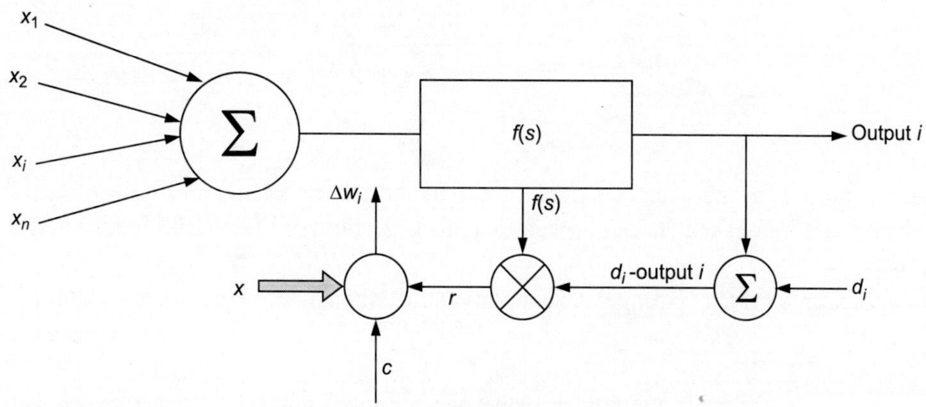

Fig. 2.10

Here the delta function is the derivative of `the activation function $f(s)$. If error E is defined as the mean squared value of $(d_i - y_i)$ given by, taking output $i = b_i$.

$$E = \frac{1}{2}(d_i - b_i)^2, \text{ then the error gradient vector.}$$

$$\nabla E = -(d_i - b_i)^2 f'(s_i)\mathbf{x} \quad \text{...(2.8.7)}$$

$$\frac{dE}{dw_{ij}} = -(d_i - b_i) f'(w_i^T x) x_j \quad \text{...(2.8.9)}$$

2.8.4 and 2.8.5 Analysis of Feed forward Network

1. The input layer (zeroth layer) redistributes the input values and does no processing.
2. The outputs of this layer are described mathematically as

$$O_i^0 = x_i \text{ where } i = 1, \ldots N^0$$
$$x_i = \text{input vector}$$
$$N^0 = \text{number of neurons in the input or zeroth layer.}$$

3. The input to each neuron in the first hidden layer in the network is a summation of all weighted connections between the input or zeroth layer and the neuron the first hidden layer. This weighted sum is called the net stimulus or net input (denoted by net).
4. The net input to a neuron from the first layer can be written as the product of input vector x_i and weight factor ω_i plus bias term.

$$\text{Net stimulus} = \sum_{i=1}^{N} \omega_i x_i + \text{bias}$$

where N represents the number of neuron in the input layer.

$$O = f(\text{net}) = \int\left[\sum_{i=1}^{N} \omega_i x_i + \text{bias}\right]$$

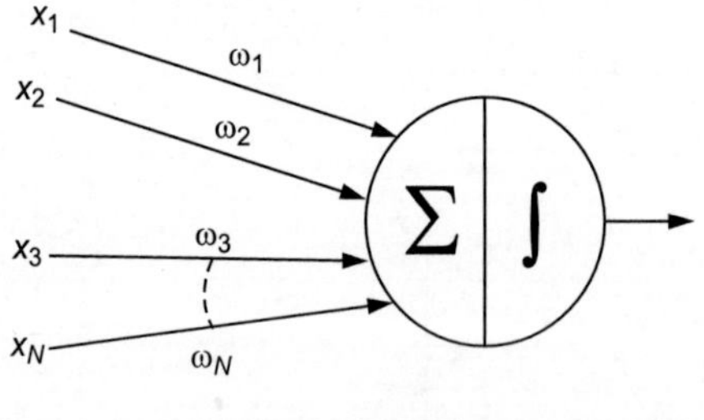

5. The net stimulus to the neuron is transformed by the neuron's activation or transfer function to produce a new value. With back propagation, this is sigmoid or linear. Bias term is added to offset the input.
6. Bias is usually a weight coming from a unitary valued input and is denoted ω_0. The final output of the neuron

$$\text{Output} = \int(\text{net}) = \int\left(\sum_{i=1}^{N} \omega_i x_i + \omega_0\right) = \int\left(\sum_{i=1}^{N_0} \omega_i O_i^0 + \omega_0\right)$$

7. Expanding to multiple neurons and multiple layers, we can see that outputs of neurons in one layer O_i^l are interconnected via weights to neurons in the next layer. So the net stimulus of jth neuron in the lth layer is given by

$$\text{net}_j^l = \sum_{i=1}^{l-1} \omega_{ji}^l O_i^{l-1} + \omega_{j0}^l$$

$$O_j^l = f_j^l\left(\text{net}_j^l\right)$$

8. The outputs which emerge from the network at the final L^{th} layer are described by

$$y_j = O_i^L \text{ where } j = 1, \ldots N^L$$

A common notation for describing the network topology (distribution of neurons in the network) of a feed forward network

$$N^0 - N^1 - N^2 \ldots N^L$$

9. When an input vector is presented to the F/F nature, the output is computed by a squared error. When we input target vectors $\boldsymbol{x}$ and target output vectors $\boldsymbol{t}$ the squared error is

$$E_p = \frac{1}{2}\sum_{j=1}^{N_L}(t_{pj} - y_{pj})^2 \quad P = \rightarrow 1 \ldots P.$$

P = number of patterns in the training set P (pattern index).

10. The output error of all vectors presented to the feed forward network is

$$E - \sum_{p=1}^{P} E_p$$

11. To reduce the error, the gradient descent error minimisation is defined as:

$$\frac{\partial E}{\partial \omega_{jk}^l} = 0$$

12. Partial derivative of total error can be broken up into a summation of the errors of each presented pattern

$$\frac{\partial E}{\partial \omega_{ji}^l} = \sum_{p=1}^{P} \frac{\partial E_p}{\partial \omega_{ji}^l}$$

13. $$\frac{\partial E_p}{\partial \omega_{ji}^l} = \frac{\partial E_p}{\partial \,\text{net}_{pj}^l} \frac{\partial \text{net}_{Pj}^l}{\partial \omega_{ji}^l}$$

The first term is the charge in error with respect to net stimulus. The second term contains the change in net stimulus with respect to weights. From the expression of net stimulus as a function of a change in weights:

$$\frac{\partial \text{net}_{pj}}{\partial \omega_{ji}^l} = \frac{\partial}{\partial \omega_{ji}^l} \sum_{k=1}^{Nl}\left(\omega_{jk}^l \, O_{pk}^{l-1}\right) + \omega_{j0}^l = O_{pk}^{l-1}$$

14. Going back to the 13th step $\dfrac{\partial E_p}{\partial \omega_{ji}^l} = \dfrac{\partial E_p}{\partial \text{net}_{pj}^l} O_{pi}^{l-1}$

This is similar to the delta rule which was used in perception learning and was a fore runner of back propagation.

15. Delta factor can be expressed as

$$\delta^l_{pj} = -\frac{\partial E_p}{\partial \text{net}^l_{pj}}$$

So,
$$\frac{\partial E_p}{\partial \omega^l_{ji}} = -\delta^l_{pj}\, O^{l-1}_{pi}$$

By defining the change with respect to the net stimulus as a delta term and substituting it back the generalized delta rule is formed because of the similarities with the delta rule developed by Widow and Hoff in the late fifties for use in training perceptrons with the least mean square method [widow 1960].

2.8.6 Adaline With a Teacher

Referring to the Fig 2.11, we have the i/p to neuron as given by $x = [x_0, x_1 \ldots x_n]$

x_0 = Bias = 1

Weight vector, $\boldsymbol{\omega} = [w_0, \ldots w_n]^T$

y_{in} is the o/p of the neuron before the non-linearity function

$$y = \text{Sgn}\ (y_{in})$$
$$= \text{Sgn}\ (\boldsymbol{\omega}^T \boldsymbol{x})$$

Where
$$t = \text{target signal}$$
$$\delta = t - y_{in} = \text{output error}$$

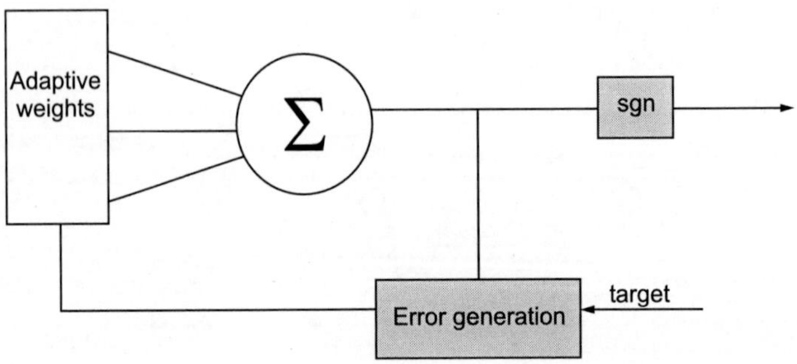

Fig. 2.11

2.8.7 Adaline Learning Algorithm

1. Assign random weight values from – 1 to + 1.
2. Set the activation inputs $x_0 = 1$.
3. Compute the net input to the neuron $y_{\text{in}} = \Sigma x_i\, w_i$.
4. Update the bias and the weights with

$$\boldsymbol{\omega}_{i\ \text{new}} = \boldsymbol{\omega}_{i\ \text{old}} + \alpha(t - y_{in})\, x_i$$

2.8.8 Back Propagation Network

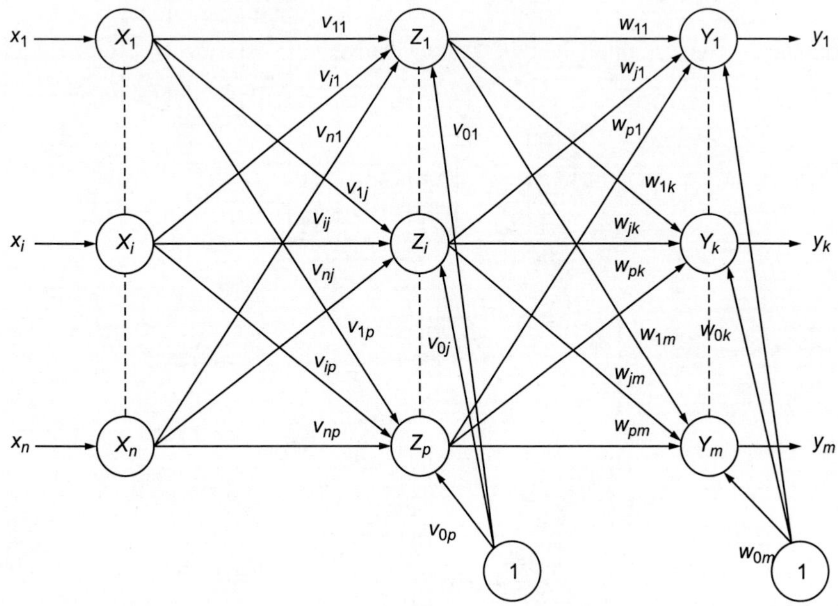

Fig. 2.12 *Back propagation Network Example*

The back propagation technique takes place in three stages.

1. Feed forward of i/p training.
2. Back propagation of associated error.
3. Weight adjustment.

Terminology

Basically we have to calculate the delta terms for the p_{output} layer first and then go backward from the output saga to the input layer, calculating deltas for the hidden neurons. The error term is found directly in the delta term for the output lags. Thus the error is propagating backward through the network. This mechanism gives rises to the name back propagation of error.

If the error E is defined as the mean squared value of $(d_i - y_i)$ given by

$E = (d_i - y_i)^2/2$,then the error gradient vector is $\Delta E = -(d_i - y_i)\ f'(s)\ x$

$$\Rightarrow \quad \frac{\partial E}{\partial W_{ij}} = -(d_i - y_i)\, f'(w_i^T\ x)\, x_j \qquad \text{...(2.8.26)}$$

The minimization of error implies that weight changes are to be in the negative gradient direction

$$w_i = -(\text{positive constant})\ \Delta E.$$

EXAMPLE 2.3: Using $w_i^{k+1} = w_i^k + \frac{c^k}{2}(d_i^k - O_i^k)y^k$, train the network of three perceptions:

Let the randomly chosen initial weight vectors be $w_1^1 = \begin{bmatrix} 1 \\ -2 \\ 0 \end{bmatrix} w_2^1 = \begin{bmatrix} 0 \\ -1 \\ 2 \end{bmatrix} w_3^1 = \begin{bmatrix} 1 \\ 2 \\ -1 \end{bmatrix}$

The values of y are: $y_1 = y_1 = \begin{bmatrix} 10 \\ 2 \\ -1 \end{bmatrix}, y_2 = \begin{bmatrix} 2 \\ -5 \\ -1 \end{bmatrix}, y_3 = \begin{bmatrix} -5 \\ 5 \\ -1 \end{bmatrix}$

Step 1: Input is y_1

$$\text{sgn}\left([1 \quad -2 \quad 0] \begin{bmatrix} 10 \\ 2 \\ -1 \end{bmatrix} \right) = \text{sgn}(10 - 4 + 0) = 1$$

$$\text{sgn}\left([0 \quad -1 \quad 2] \begin{bmatrix} 10 \\ 2 \\ -1 \end{bmatrix} \right) = \text{sgn}(0 - 2 - 2) = -1$$

$$\text{sgn}\left([1 \quad 3 \quad -1] \begin{bmatrix} 10 \\ 2 \\ -1 \end{bmatrix} \right) = \text{sgn}(10 + 6 + 1) = 1$$

EXAMPLE 2.4: A two layer feed forward network using neurons having bipolar binary activation function the output O_5 has to be determined and the region for $O_5 = 1$ is to 0.

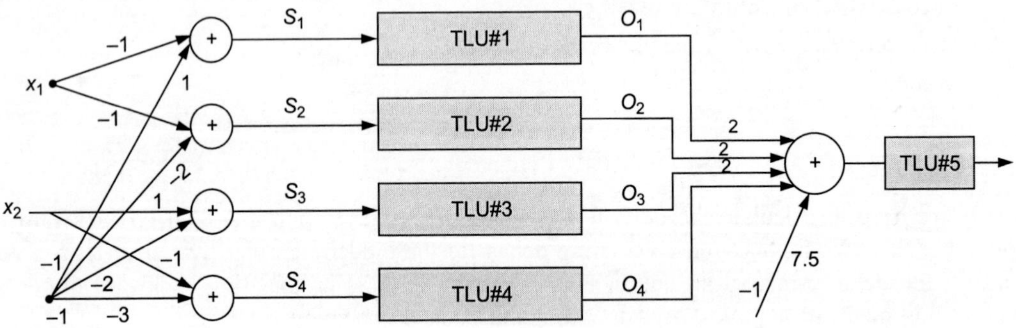

The input vector $\begin{bmatrix} x_1 \\ x_2 \\ -1 \end{bmatrix}$ or $[x_1 \quad x_2 \quad -1]^T$

The output vector is $[O_1\ O_2\ O_3\ O_4]^T$

$O_1 = \text{sgn}(s_1) = \text{sgn}(x_1 - 1)$ as threshold units use sgn function.

$O_2 = \text{sgn}(s_2) = \text{sgn}(-x_1 + 2)$

$O_3 = \text{sgn}(s_3) = \text{sgn}\ (s_2)$

$O_4 = \text{sgn}(s_4) = \text{sgn}(-x_2 + 3)$

$O_5 = \text{sgn}(2O_1 + 2O_2 + 2O_3 + 2O_4 - 7.5)$

For O_5 to be 1, the only condition suitable that bracketed expression must be +ve which implies that $O_1 = O_2 = O_3 = O_4 = 1$.

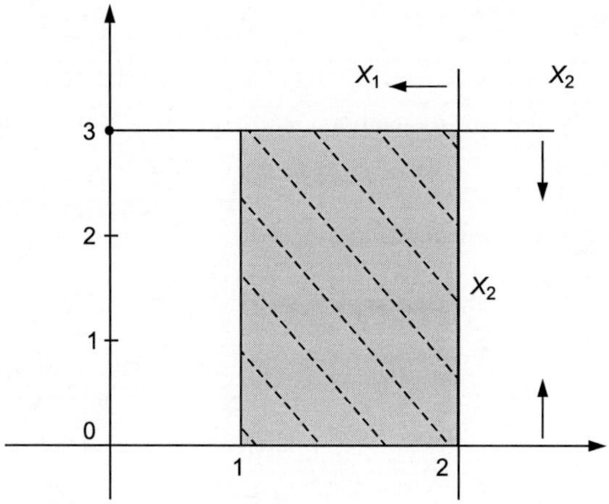

Fig. 2.14

If we plot the outputs in X_1, X_2 coordinates, it is evident that for O_1 to be 1, $x_1 - 1 > 0$, i.e., X_1 should be greater than 1. For O_2 to be 1, $(-x_1 + 2) > 0$, i.e., X_1 should be less than 2. For O_3^- to be 1 $x_2 > 0$ for O_4 to be 1, $(-x_2 + 3) > 0$, so x_2 should less then 3. The hatched section shows the region whose $O_5 = 1$. Outside this region $O_5 = -1$. Hence this network can be made to work as a classifies for consideration of $O_5 = 1$ and -1.

2.8.9 Correlation Learning Rule

This is a special case when $r = d_i$ in the equation

$$\Delta \mathbf{w}_i = c\, d_i \mathbf{x} \qquad \text{...(2.8.27)}$$

The adjustment can be made by

for $\quad j = 1, 2, 3 \ldots n$ as $\Delta w_{ij} = c\, d_i x_j$

By this rule, the weight increase is given by the product of desired response and input. This can be thought of as a special case of Hebbian rule with a binary activation function. Unlike Hebbian rule which works for unsupervised environment, the correlation learning is a case of supervised type as we are involving the desired response d_i. The weight initialization is like Hebbian ($\mathbf{w} = 0$).

2.8.10 Winner-Take-All Learning Rule

This is case of competitive learning in unsupervised mode. Figure 2.15 the premise is that one of the neurons in a layer has the maximum response; it is called the winner. If the corresponding winning neuron has the weights (and is in m^{th} layer).

$$\mathbf{w}_m = [w_{m1}, w_{m2}, \ldots, w_{mn}]^T \qquad \text{...(2.8.29)}$$

The increment is computed by $\Delta \boldsymbol{w}_m = \alpha(\boldsymbol{x} - \boldsymbol{w}_m)$ for m^{th} layer neuron having maximum response.

$$\Delta w_{mj} = \alpha(x_j - w_{mj}) \qquad \ldots(2.8.30)$$

for $\qquad j = 1,2,3 \; \ldots, \; n$

Where $\alpha > 0$ is a small learning constant which decreases with progress of learning. If there are p neurons participating in the competition.

$$\boldsymbol{w}_m^T x = \max\; (\boldsymbol{w}_i^T \boldsymbol{x}) \; i = 1, 2, \ldots, p$$

It may be noted that only the fan-in weight vector of the winning neuron is adjusted. Winning neighborhood can be extended.

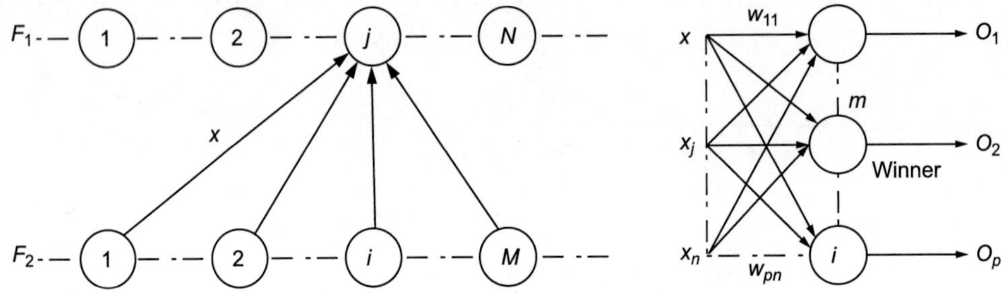

Fig. 2.15

2.8.11 Outstar Learning Rule

This rule is related to supervised learning and allows the network to utilize statistical properties of input and output signals. This is designed to have desired response of the layers of p neurons here,

$$\Delta \boldsymbol{w}_j = \beta(d - \boldsymbol{w}_j) \qquad m = 1, 2, \ldots, p$$

$$\Delta w_{mj} = \beta(d_m - w_{mj}) \quad m = 1, 2, \ldots, p$$

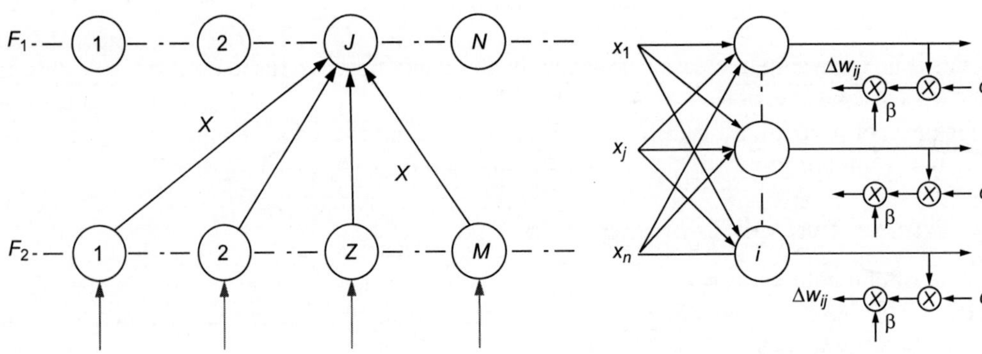

Fig. 2.16

2.8.12 Widrow-Hoff Learning Rule

This is applicable for supervised training. It minimise the squared error between the desired output d_i and neuron's activation value.

The weight vector increment is given by

$$\Delta w_i = c\,(d_i - \mathbf{w}_i^T\mathbf{x})\,\mathbf{x}$$

For single weight adjustment we have:

$$\Delta w_{ij} = c\,(d_i - \mathbf{w}_i^T\mathbf{x})\,x_j$$

This rule is a special case of delta learning rule. It is also called LMS (Least Mean Square) learning rule.

Learning may be defined as a relatively permanent change in behaviour which is caused by practice or experience. The relative permanent change in behaviour is also called self-organisation or adaptation. Learning is described as relatively permanent, rather than absolutely permanent, to allow space for forgetting.

Learning in biological system occurs when effective coupling existing between the neurons is modified. Similarly learning in ANNs can be considered to occur when the weight of the input links are changed.

Hebbian learning law is powerful yet simple *j* however, the law needs some modifications to take into account some practical aspects such as:

(*a*) Biological systems forget as well as learn
(*b*) There are excitary and inhibitory inputs and
(*c*) A limit has to exist on the increase of reinforced input.

In Neo-hebbian learning (Grossberg) biological forgetting is incorporated. In Kohonen's learning model, neurons compete with each other in learning. The neuron with largest output wins. Its neighbours are allowed to adjust the weights of their inputs whereas the competing neurons are suppressed.

In back propagation learning, weights are set at random in the beginning. Learning takes place in two phases: forward and backward.

The error between the actual output and desired output is calculated and it is propagated backwards to adjust weights.

In Fuzzy learning expert knowledge represented by of then rules is used thus model takes neural networks a step closer to the biological system.

2.9 APPLICATIONS OF ANN'S

1. Pattern recognition and image processing

   (*a*) Recognition of visual images
   (*b*) Recognition of hand written characters
   (*c*) Recognition of printed characters
   (*d*) Speech and other pattern recognition based tasks.

2. Optimization and constraint satisfaction:

   (*a*) Problems having constraints like obtaining optimal solutions for manufacturing schedule.
   (*b*) Finding shortest possible time frame.

3. Forecasting and Risk Assessment:
   (*a*) Stock market banking and Econometrics.
   (*b*) Weather forecasting in Meteorology.
4. Control systems:
   Very popular in Japanese systems where control of plants and robotics invariably use neural networks.

Algorithm

Initially all the weights and bias are set to 0.1 then we can present the input pattern to be classified. At the input layer, the activation function used is identity; hence the output from the input layer remains same as the input presented. Also, the activation for the output unit is also set. Then the weights are updated based on the Hebb's learning rule as in section an epoch is completed after presenting all the samples of the input pattern. The stepwise algorithm to train Hebb's net is as follows.

Step 1: Initialize all weights and bias to zero. $W_i = 0$ for $i = 1$ to n. Where n is the number of input neurons.

Step 2: For each input training vector and target output pair (S, t) perform Steps 3 – 6.

Step 3: Set activation for output units with input vector. $X_i = s_i$ $(i = 1$ to $n)$

Step 4: Set activation for output unit with the output neuron $Y = t$.

Step 5: Adjust the weights.

Step 1: Initialize weights, not zero but small random values are used. Set learning rate α.

Step 2: While stopping condition is false, do Steps 3-7.

Step 3: For each bipolar training pair $s : t$, perform Steps 4-6.

Step 4: Set activations of input unit $x_i = S_i$, for $i = 1$ to n.

Step 5: Compute net input to output unit $y_{-in} = b + \Sigma_i W_i X_i$

2.10 SUMMARY OF LEARNING RULES

L. Rule	Δw_{ij} $j = 1, 2, \ldots, n$	*Type Learning*	*Initial Weights*	*Neuron Characteristic*
1. Hebbian	$cf(s_i)x_j$	unsupervised	O	Any
2. Perception	$c[d_i - \text{sgn}(w_i^T x)]x_j$	supervised	random	binary or unipolar/bipolar
3. Delta	$c(d - f(s_i)f'(s_i)x_j$	supervised	random	continuous
4. Widrow-hoff	$c(d_i - w_i^T x)x_j$	supervised	random	Any
5. Correlation	$cd_i x_j$	supervised	O	Any
6. Winner-take-all	$\Delta w_{mj} = \alpha(x_j - w_{mj})$	unsupervised	random	continuous
7. Outstar	$\beta(d_i - w_{ij})$	supervised	O	continuous

SUMMARY

- This unit explains the basics of ANN operation Activation function and architecture have been discussed.
- Learning strategy has been divided into supervised, unsupervised and reinforcement types.
- Rules of learning have been introduced including generalized delta rule and back propagation.
- Suitable examples have been selected to explain the rules.
- Special remarks on learning are included to broaden the base of graph of the students.

Short Questions

1. What is sigmoid function?
2. Draw the sketches for any continuous function?
3. State the Hebbian postulate?
4. What is LMS rule?
5. Is correlation rule supervised type?
6. What is a learning constant?
7. What is the meaning of back propagation?
8. Compare Preceptron rule and Delta rule?
9. Give three parameters of brain for neural dynamics?
10. Explain the importance of linear separating?

Long answer Questions

1. **Explain differential neuron Activation function?**
   Refer 2.3
2. **Discuss learning strategy with necessary example. Write about supervised and unsupervised learning rule with examples?**
   Refer 2.7 + 2.7.1 + 2.7.2 + 2.7.5 + 2.7.6
3. **Discuss the operation of ANNs?**
   Refer 2.2
4. **Write a short notes on neural dynamics?**
   Refer 2.6
5. **Discuss architecture of ANN?**
   Refer 2.4
6. **Compare supervised and unsupervised learning Rule?**
   Refer 2.7.8
7. **Discuss different types of learning rue with details?**
   Refer
8. **Using Hebbian learning rule find out the weight vector to train the following input vectors. $X_1 = [1 \;\; -2 \;\; 1.5 \;\; 0]$, $X_2 = [1 \;\; -0.5 \;\; -2 \;\; -1.5]$, $X_3 = [0 \;\; 1 \;\; -1 \;\; 1 \;\; -5]$. Assume initial weight vector to be $w_1 = [1 \;\; -1 \;\; 0 \;\; 0.5]^T$.**
   Refer Example 2.1

9. Continuous activation functions are used with $\lambda = 1$, Output 1 = 0.3, output 2 = –0.5. Find the input vector applied.

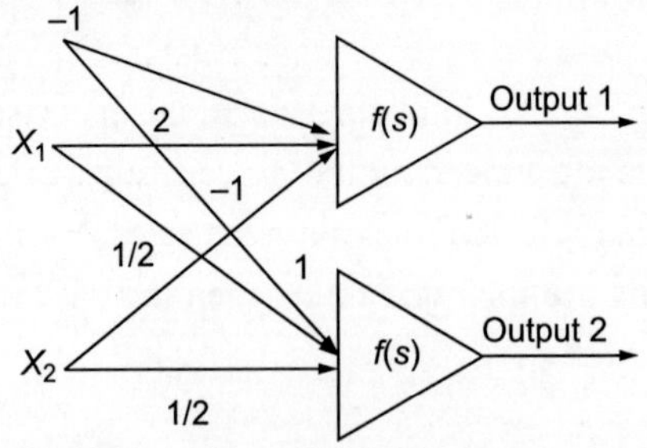

10. Assume that vertices of a three dimensional bipolar binary cube are used to repeated 8 states of recurrent network with 3 bipolar binary neurons. The equilibrium states are 111 and –1–1–1. Sketch the desirable state transitions.

❑❑❑

3

SINGLE LAYER FEED FORWARD NEURAL NETWORKS

3.0 FOREWORD

"Perceptron" originates from perception or perceive. ROSENBLATT's perceptron was basically a device meant for pattern recognition. A basic perceptron can be developed on the principles of Mc Culloch and Pitts model of neuron. Logic gates developed on single perceptron solved many cases but failed in the case of EX-XOR. This led to hue and cry in the scientific circles when Minsky and Papert wrote against neural Logic. In fact this was the singular set back for the neural networks in the early years of development.

In this unit you will be made familiar with Perceptron's role as classifier based on discriminant functions decision surfaces. The role of discrete and continuous Perceptron, non-parametric training methods will be examined with the purpose of development of multi-category Perceptron's. All these have a great impact on the study of pattern recognition techniques. The concepts of classification have been clarified in an additional section.

Linear separability concepts are integral part of the basic Perceptron working philosophy. These ideas have led to formulation of the famous Perceptron convergence theorem based on Cauchy-Schwartz inequalities. This theorem implies that of linear separability is feasible then there must be finite number of steps of iteration in case of adjustments of the weights in the course of training procedure. The finite steps stress on the importance of convergence of the training procedure. Limitation of Perceptron models and application are taken up at the end of the unit.

3.1 INTRODUCTION

Basically Perceptron is a computational element modelled on the retina of the eye. It draws its name because of this background. In the original model of Perceptron suggested by Rosenblatt, the sensory unit consists of photo detectors which process the incoming images into digital (0/1) signal. If the input signals exceed a certain threshold, the output is 1. The photo detectors are randomly connected (Fig. 3.1) to the association unit which has predicates responsible for examining the output of sensory unit for specific images. The next unit is recognizer unit having perceptron for recognizing the patterns. The weights of S and A are fixed where as these of *R* are adjustable. The output *R* is such that if the weighted sum is less than or equal to zero, the output is zero or equal to the weighted sum.

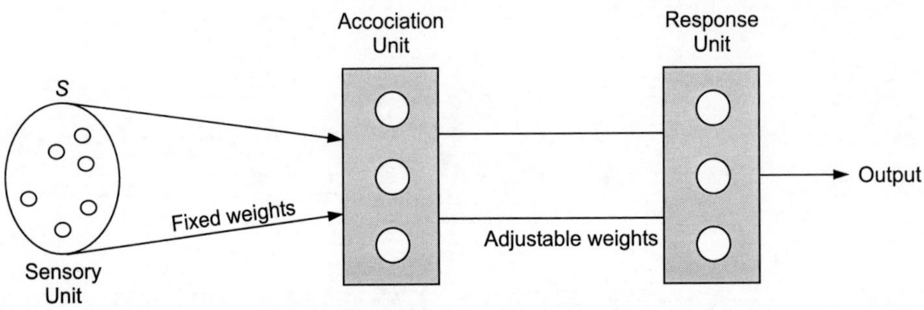

Fig. 3.1 *Perceptron of Rosenblitt.*

One of the most important uses of Perceptron's is to classify. As is known, electronic neural networks are inspired by the biological neural system. Not all the neural models have biological justification. A major function of neural network is classification having application in pattern recognition, communications and control. Several possible classification schemes exist. "It depends on who is doing the classification and what point of view is to be conveyed".

3.1.1 Pictorial Illustration of Classification by Perceptron

Consider a two dimensional case of 2-input Logic OR function. The output is one if the input patterns are 01, 10, and 11. The output is zero for the pattern 00.

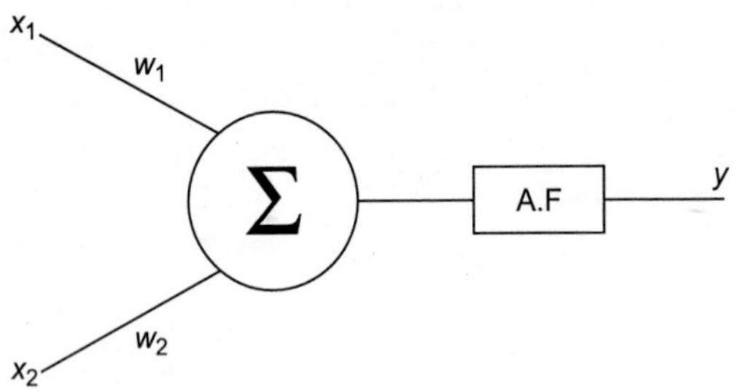

x_1	x_2	y
1	1	1
0	1	1
1	0	1
0	0	0

Fig. 3.2 *Pictorial Illustration of Classification by Perceptron with Truth Table.*

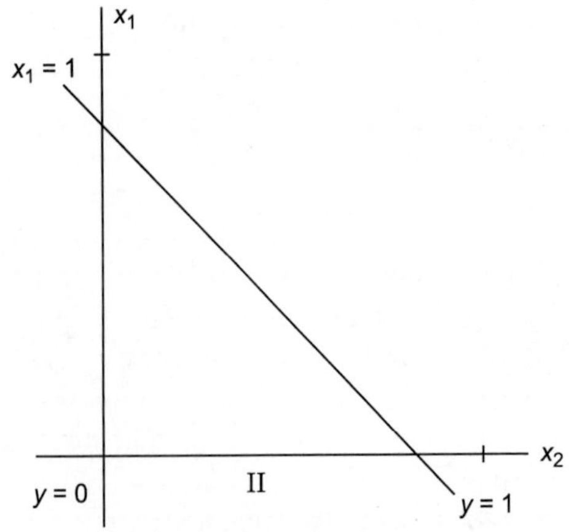

Fig. 3.3 *Line Representation.*

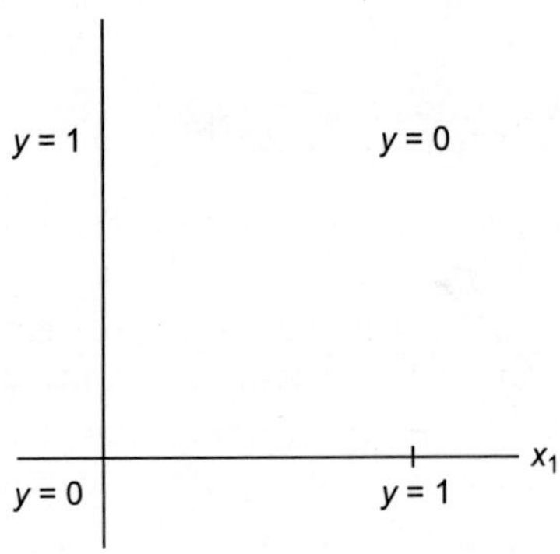

Fig. 3.4 *XOR representation by graphical Method.*

For this Perceptron, a line L can be found such that the space is divided for class I ($y = 1$) and class II ($y = 0$). The basic objective of a training algorithm is to adjust the weights. w_1 and w_2 such that the perceptron produces the line of separation. Not all patterns are linearly separable for a perceptron. A classical example is the case of XOR Logic function. (Fig. 3.3)

There is no linear division of space to produce the classification. We can say that single perceptron cannot solve this problem. (multiple perceptron do offer solution for XOR)

The operation of a neural processing element can be viewed as a non-linear mapping operation. The mapping converts the input vector to a scalar. This mapping can be considered to occur in two places:

(*a*) Linear weighted mapping to followed by '*s*'

(*b*) Non-linear mapping of s to y through a function f^n. $y = f(s) = f(\mathbf{w}.\ \mathbf{x})$

In general $y = f(w © x)$ where © is Confluence operation (scalar or distance measure).

3.1.2 Example of Classification

The problem is to have an algorithm to train a perceptron to classify hand written letters such A or B.

Sol: Let the perceptron output be 1 when letter in detected as A and 0 when it is B. The algorithm can be:

- As sign random values to the weights and threshold
- Present an input (or digitized picture)
- Calculate the output $y = f\left[\sum_{i=1}^{n} w_i(t)\, x_i(t) - \theta\right]$
- Adjust the weights and reduce the error
- If y is correct $w_i(t + 1) = w_i(t)$; if $y = 0$ and should be 1 $w_i(t + 1) = w_i(t) + \alpha\, x_2(t)$
  If $y = 1$ and should be 0 $w_i(t + 1) = w_i(t) - \alpha\, x_i(t)$.
  α is a positive fraction which is controls the adaption rate.

The above discussed perceptron algorithm can be utilized to develop the concepts of decision boundary in the context of classification.

The points of consideration are:

1. Each shape represents objects, which belong to same set or class.
   For example, A circle can represent all men and ellipse can represent the birds.
2. The neuron (in the mode of perceptron) can find a line, which separates two classes.
3. The lines represent the locus of the neuron activation values possible for a given set of weights associated with the neuron.
4. For a particular case with weighted inputs, the activation function must lie on the line.
5. The decision machines (or) decision engines, which follow this type of equation, are called linear classifiers.
6. By changing w_0, w_1, w_2 and the net stimulus value, the position of the line can be moved until one side of line contains examples of a given class and the other side contains the other class. The process of change w_0, w_1, w_2 is a part of training program.
7. When more than two dimensions are involved, the equation becomes a representation of a hyper plane of separation.

8. The data involved in the above process are called linearly separable data as compared to non-linearly separable data.
9. A perceptron has a special algorithm for linear separation. It adjusts the weights using the error vector between data point and a separating decision line or boundary.
10. Change in weights are made as per the training program until none of the trained data produces error. The Perceptron learning rule requires that a desired output is known before the network can adopt the weights to find the line which separates one class from the other. This process of using desired outputs for training the neural network is known as supervised learning.
11. The supervised learning employs a teacher to assist the network by telling the network what the desired response to a given stimulus is to be. The operation is similar to a student being guided by a teacher.
12. Learning system is exposed to the environment which is represented by measurement vector (MV) of features MV is also presented to a teacher who by experience determines the desired response.
13. This desired response is used to create error signal which adapts weights of the learning system.
14. Each input vector has an associated desired output vector which is used for the training.
15. In case of supervised learning, every input vector has a corresponding desired output. (This is not practicable in all cases).

A basic representation of such concepts can be modelled in Fig. 3.5.

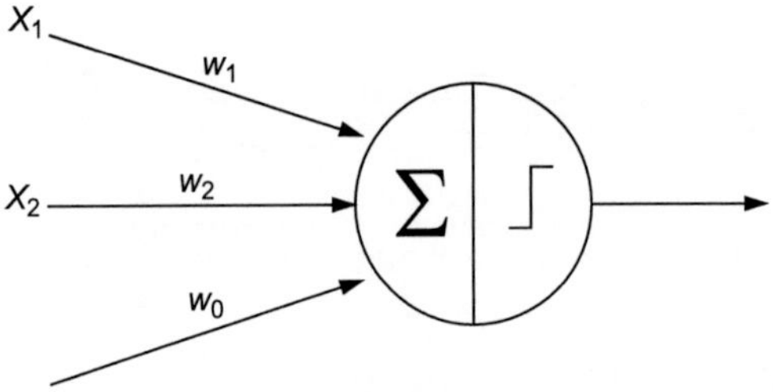

Fig. 3.5

N_s = Net stimulus = $w_0 + w_1x_1 + w_1x_1 = 0$ for the decision line.

$$\Rightarrow \qquad x_2 = -\frac{w_1x_1}{w_2} + \frac{w_0 - N_S}{w_2} \qquad ...(3.1.1)$$

$$w_0 \rightarrow \text{bias factor}$$

the above equation represents a straight line.

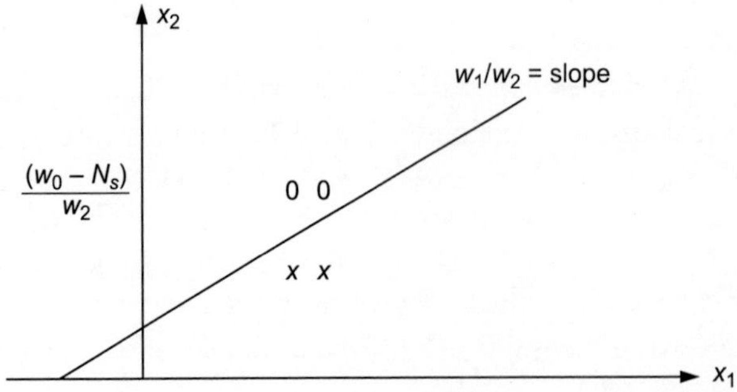

Fig. 3.6 *Linear Classifier.*

3.2 PERCEPTRON MODEL

The models depicted in Fig. 3.7 and 3.8 may be called discrete (binary) and continuous perceptrons. Rosenblatt introduced discrete perceptron as the first learning machine. In this, we have

(*a*) Synaptic weights
(*b*) Summation performed by summing Node
(*c*) Hard limiting threshold operation performed by Threshold Logic unit.

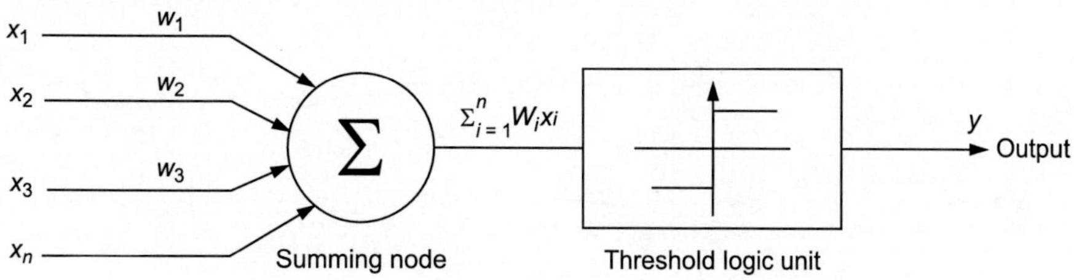

Fig. 3.7 *Hard Limiting Discrete (Binary) Perceptron.*

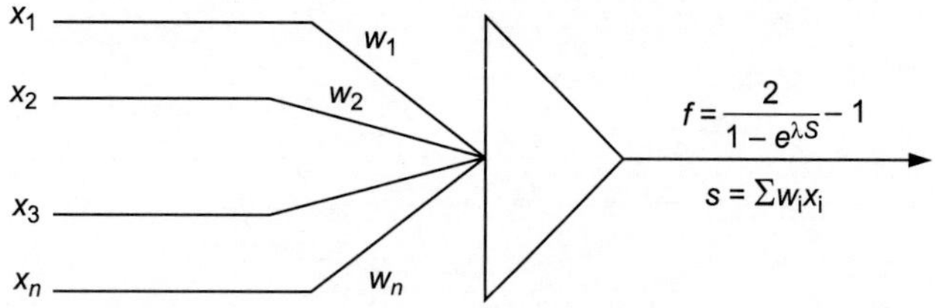

Fig. 3.8 *Continuous Perceptron.*

In case of continuous perceptron, the function f is differentiable.

For $$f(s) = \frac{1}{1+\bar{e}^{\lambda s}} \quad f'(s) = \lambda\, f(s)\,[1 - f(s)]$$

Proof: $$\text{LHS} = f'(s) = \lambda(1+\bar{e}^{\lambda s})^{-2}(\bar{e}^{\lambda s}) = \frac{\lambda e^{-\lambda s}}{(1+e^{-\lambda s})^2}$$

$$\text{RHS} = \lambda\ f(s)\ [1 - f(s)]$$

$$= \frac{\lambda}{1+\bar{e}^{\lambda s}}\left[1 - \frac{1}{1+\bar{e}^{\lambda s}}\right] = \frac{\lambda}{1+\bar{e}^{\lambda s}}\left\{\frac{1+\bar{e}^{\lambda s} - 1}{1+\bar{e}^{\lambda s}}\right\} = \frac{\lambda \bar{e}^{\lambda s}}{(1+\bar{e}^{\lambda S})^2} = f'(s) \quad ...(3.2.2)$$

This property is of significance in back propagation techniques.

Note: For discrete case the domain of the vectors are defined in m-dimensional space for $i = 1, 2....m$

$$(-1,\ 1)^m = \{y \in R^m,\ y_i \in (-1\ 1))\} \text{ Bipolar}$$
$$(0,\ 1)^m = \{y \in R^m,\ y_i \in 0(0\ 1)\} \text{ Unipolar}$$

3.2.1 Continuous Perceptron

The weights of TLU are to converge to some solution w^{ko} for any positive correction increment constant. Using gradient (or steepest decent) technique, start is done from an arbitrary weight vector w and gradient $\Delta E(w)$ of the current error is computed. The updating is done by moving in the direction of the negative gradient along the multidimensional error surface (the direction of steepest descent is nothing but the direction of negative gradient). The basic algorithm is $w^{k+1} = w^k - \eta \Delta\ E\ w^k$. Where η is the learning constant. The expression of classification error to be minimized is

$$E_k = \frac{1}{2}(d^k - y^k)^2 \text{ or } E_k = \frac{1}{2}[d^k - f(w^{kT} x^k)]^2 \qquad ...(3.2.3)$$

The aim is to minimize the error function $E(w)$ in $(n + 1)$ dimensional weight space (Fig 3.9).

For a well behaving error function having a minimum at $w = w_f$, finite number of steps have to result in $E(w_f) = 0$. The movement of weight vector is noticeable on the error surface as well as the contour lines.

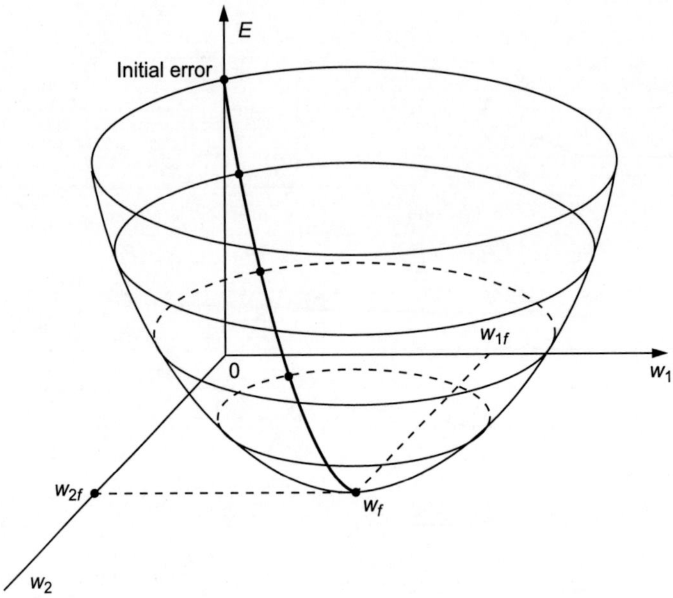

Fig. 3.9

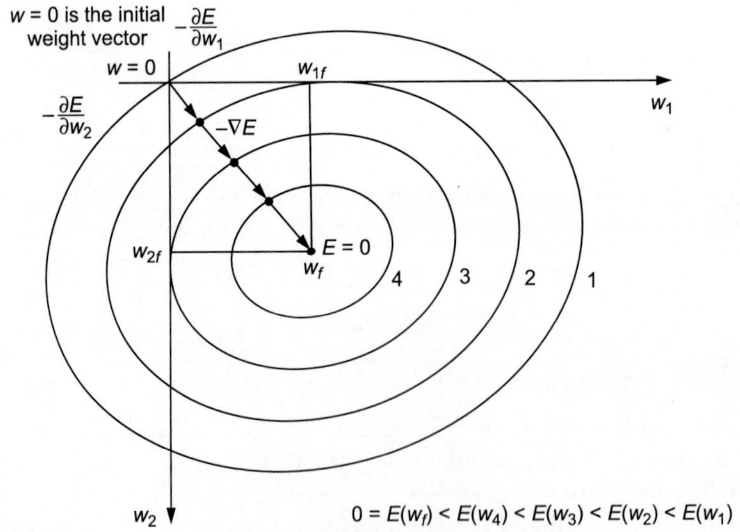

Fig. 3.10

For the sake of convenience, we can take the expression.

$$\text{As } \nabla E(w) = \begin{bmatrix} \frac{\partial E}{\partial w_1} \\ \vdots \\ \vdots \\ \frac{\partial E}{\partial w_{n+1}} \end{bmatrix} \Rightarrow \nabla E(w) = -(d-0)f'/(s) \begin{bmatrix} \frac{\partial (s)}{\partial w_1} \\ \vdots \\ \vdots \\ \frac{\partial (s)}{\partial w_{n+1}} \end{bmatrix} \quad \text{...(3.2.5)}$$

As $s = w_y^T \Rightarrow \frac{\partial (s)}{\partial w_1} = x_i \; i = 1, 2 \ldots n+1$

or $\Delta^E(w) = -(d - 0)\, f'/(s)x$ ...(3.2.6)

or $\frac{\partial E}{\partial w_i} = -(d-0)\, f'(s)x_i$. This is the training rule for the continuous perceptron.

Taking a bipolar continuous activation function $f(s)$ of the form

$$f(s) = \frac{2}{1+\exp^{(-s)}} - 1 \; f'(s) = \frac{2\exp(-s)}{[1+\exp(-s)]^2} \quad \text{...(3.2.7)}$$

As $$\frac{2\exp(-s)}{[1+\exp(-s)]^2} = \frac{1}{2}(1-o^2)y$$

$$\nabla E(w) = -\frac{1}{2}(d-0)(1-o^2)y \quad \text{...(3.2.8)}$$

This leads to the complete delta rule for the bipolar continuous activation function.

$$w^{K+1} = w^K = \frac{1}{2}\eta\,(d^K \quad o^K)(1-o_k^2)y^k \quad \text{...(3.2.9)}$$

This weight adjustment has the same direction as the discrete perceptron rule. The factor $1 - O_K^2$ Works as a moderating factor. This is always positive and smaller than 1. A noteworthy feature for discrete perceptron training algorithm is that the solution is always for linearly separable problem. The negative gradient training does not guarantee solution of linearly separable patterns.

3.3 CLASSIFICATION

To study the multi category perceptron it is prudent to have a detailed look into classifiers on the concept of discriminate function.

3.3.1 Classifiers

In a simple classifier shown if $\text{sgn}\,(x_1 + x_2 + x_3) = \begin{cases} 1 & \text{Category I} \\ -1 & \text{Category II} \end{cases}$ ...(3.3.1)

With $w_1 = w_2 = w_3 = 1$ all others being 0.Then a classification into two categories is possible through this decision function. The partitioning plane is $(x_1 + x_2 + x_3) = 0$ all the points above it are mapped into +1 and all below it are mapped into –1.

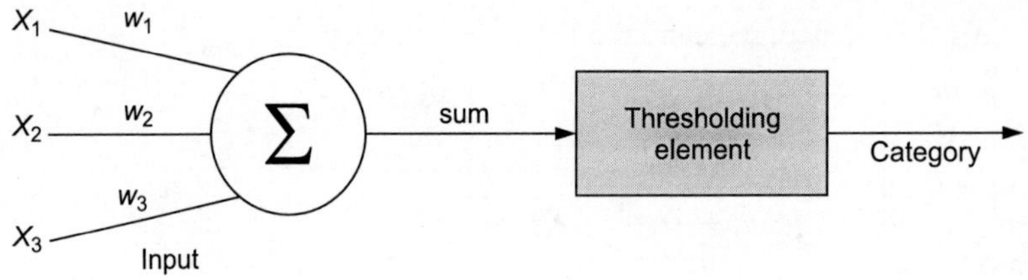

Fig. 3.11

3.3.1.1 Pattern

A pattern is the quantitative description of an object, event or phenomenon. For classifying patterns, we may use spatial or temporal types. Examples of spatial patterns can be weather maps or finger prints. Electrocardiograms signal Vs time curves are examples of temporal patterns. The basic aim of pattern classification is to assign a physical object, event or phenomenon to be put into one of pre-specified classes or categories.

3.3.1.2 Recognition and Classification

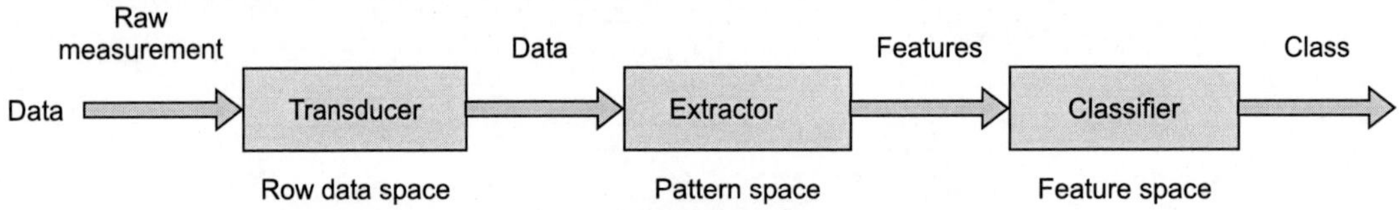

Fig. 3.12

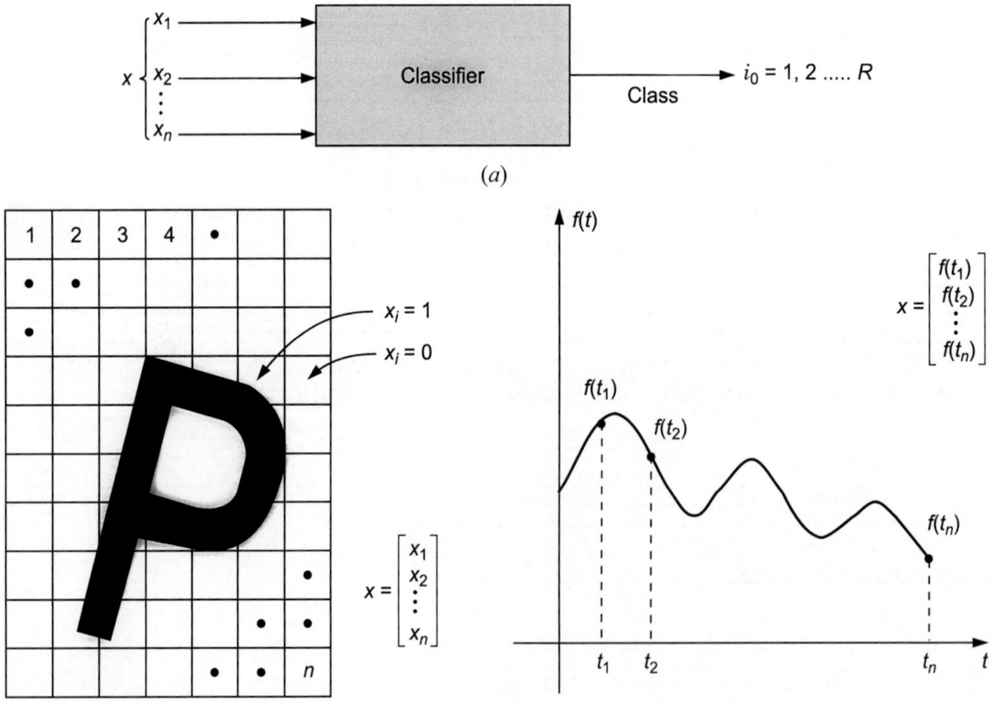

(a)

(b)

Fig. 3.13

Two simple ways of coding patterns are (*a*) spatial (*b*) temporal. Mathematically, Classification at the system's output is obtained by the implementation of decision function 'i_0' (x). The responses 'i_0' (1, 2....R) represent the categories of classification.

Geometrically, a pattern space can be represented by a point in n-dimensional Euclidean space called the pattern space. A pattern classifier maps sets of points in E^n space into one of the numbers described by the decision function. An example of $n = 2$ and $R = 3$ is shown in Fig. 3.14.

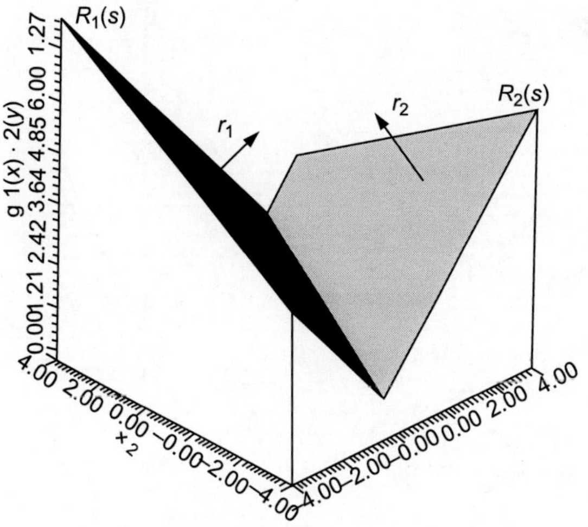

Fig. 3.14

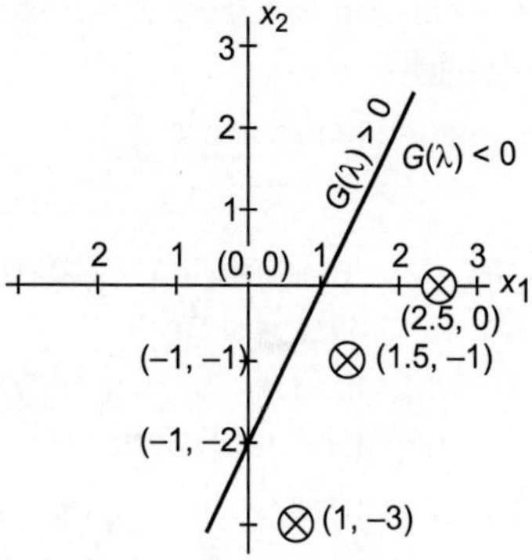

Fig. 3.15

The region denoted by X_i are called decision regions separated from each other by decision surfaces. For two dimensional cases (E^2) decision surfaces are curved lines. In a more general case E^n there may be (n – 1) dimensional hyper surface.

3.3.2 Discriminant Function

The discriminant functions $G_i(x)$ are such that the pattern vector x belongs to i^{th} category if and only if $G_i(\mathbf{x}) > G_j(x)$ for $i, j = 1,2 \ldots R, i \neq j$. R is the total number of discriminant functions.

Within the "own region" x_i, the ith discriminant will have the largest value. These are used by the classifier during the classification process to determine membership of a pattern by the method of comparison. These functions have scalar values.

The equation for decision surface is that

$$G_i(\mathbf{x}) = G_j(\mathbf{x})$$

Example: Six patterns in 2-D pattern space shown in Fig. 3.15 are to be classified according to their membership.

$[0\ 0]^T\ [-1\ -1]^T\ [-1\ -2]^T$ class 1

$[2.5\ 0]^T\ [1.5\ -1]^T\ [1\ -3]^T$ class 2

An arbitrary decision surface can be

$$\left.\begin{aligned} &\frac{2}{3}x^2 - 2x_1 + 2 = 0 \\ &2x_2 - 6x_1 + 6 = 0 \end{aligned}\right\} \text{Representing the line of separation}$$

Two contiguous decision regions x_1 x_2 are separated by the above equation where

$$G(x) = 2x_2 - 6x_1 + 6$$

3.3.3 Construction of Discriminant Function

In a case of six pattern vectors in 2-D space having values.

$[0\ 0]^T$ $[-0.5\ -1]^T$ $[-1\ -\ 2]^T$ class 1

$[3\ 0]^T$ $[1.25\ -1]^T$ $[1\ -2]^T$ class 2

The decision surface is shown in Fig. 3.15 is $G(x) = -2x_2 - x_2 + 2$

The plane equations are: $2x_1 - x_2 + 2G_1(x) - 4 = 0$

$$-2x_1 - x_2 + 2G_2(x) = 0$$

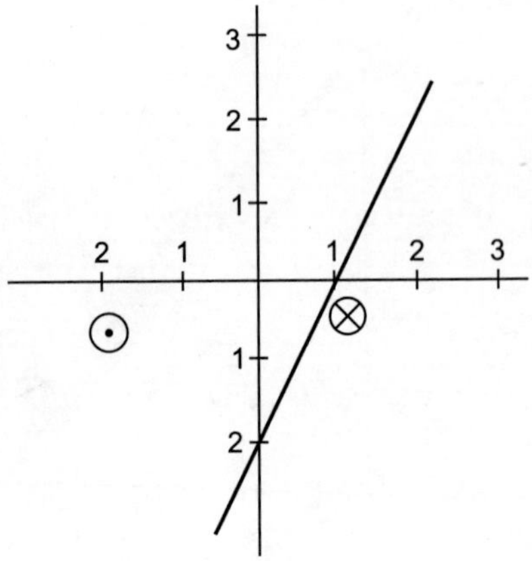

Fig. 3.15

1. The decision line has two normal vectors: they are $[2\ -1]^T$ and $[-2\ 1]^T$.

   The discriminate function $G_1(x)$ can be constructed by selecting its 3-tuple unit normal vector n_1.

2. This can be done by augmenting $[2\ -1]^T$ by a positive value say 2. The resulting normal vector is $[2\ -1\ 2)]^T$

$$n_1 = \left[\frac{2}{3}\ \frac{-1}{3}\ \frac{2}{3}\right]^T \Rightarrow n_2 = \left[\frac{-2}{3}\ \frac{1}{3}\ \frac{2}{3}\right]^T$$

3. Assume that discriminate function planes $G_1(x)$ and $G_2(x)$ intersect at $x_1 = 1$ and $x_2 = 0$. Intersection point is $[1\ 0\ 1]^T$.

$$n_1'\left(\begin{bmatrix} x_1 \\ x_2 \\ G_1 \end{bmatrix} - \begin{bmatrix} 1 \\ 0 \\ 1 \end{bmatrix}\right) = 0$$

$$n_2'\left(\begin{bmatrix} x_1 \\ x_2 \\ G_2 \end{bmatrix} - \begin{bmatrix} 1 \\ 0 \\ 1 \end{bmatrix}\right) = 0$$

so

$$G_1(x) = \left[-1\frac{1}{2}\right]\begin{bmatrix} x_1 \\ x_2 \end{bmatrix} + 2$$

$$G_1(x) = \left[1\frac{1}{2}\right]\begin{bmatrix} x_1 \\ x_2 \end{bmatrix}$$

3.3.4 Dichotomier

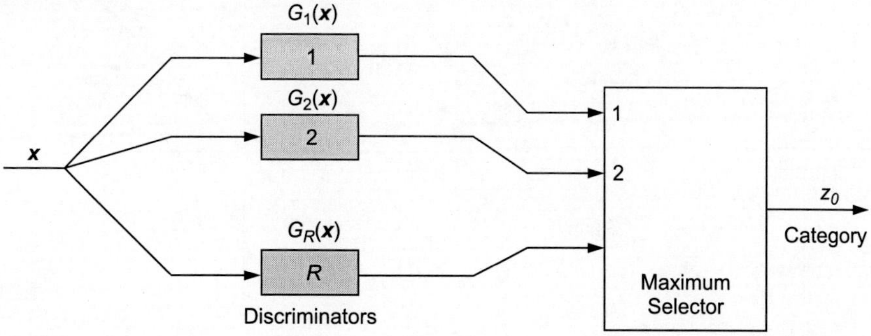

Fig. 3.17

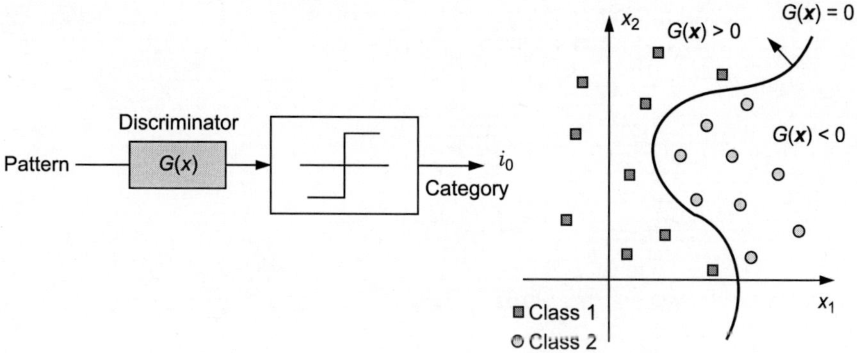

Fig. 3.18

The classifier for the special case of $R = 2$ is called Dichotomier.

$$i_0 = \text{sgn}[G(\boldsymbol{x})] = \begin{cases} +1 & G(\boldsymbol{x}) > 0 \\ \text{undefined} & G(\boldsymbol{x}) = 0 \\ -1 & G(\boldsymbol{x}) < 0 \end{cases}$$

3.3.5 Analysis

3.3.5.1 *Linear Classifier*

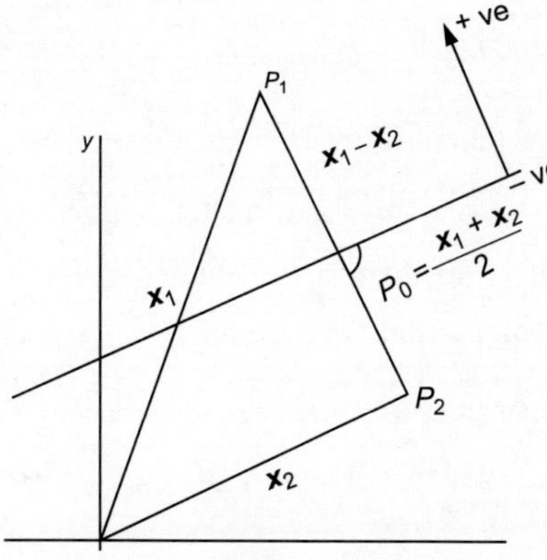

Fig. 3.18

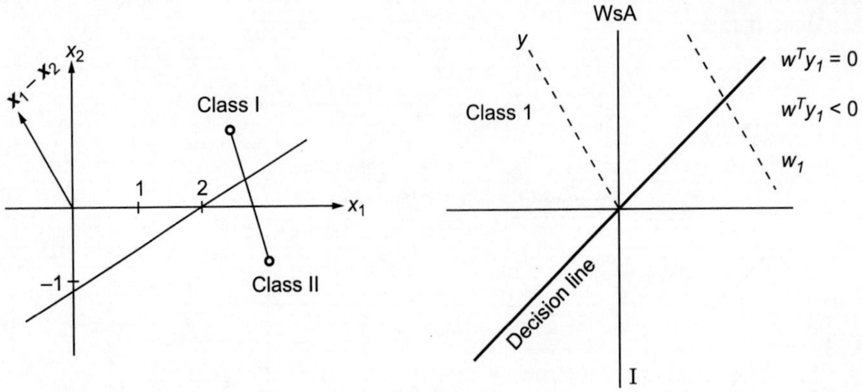

Fig. 3.19

In the normal vector form (Fig. 3.18)
$(x_1 - x_2)^T (x - x_0) = 0$ for the line which is normal to vector $(x_1 - x_2)$
And centre point P_0

$$\Rightarrow \quad (x_1 - x_2)^T - \frac{1}{2}(x_1 - x_2)^T (x_1 - x_2) = 0 \quad \text{...(3.3.3)}$$

as P_0 is $\dfrac{x_1 + x_2}{2}$

So $$(x_1 - x_2)^T x + \frac{1}{2}(x_2^T x_2 - x_1^T x_1) = 0$$

As we have $x^T x = \|x\|^2$, we can rewrite the above

$$(x_1 - x_2)^T x + \frac{1}{2}(\| x_2 \|^2 - \| x_1 \|^2) = 0 \quad \text{...(3.3.5)}$$

This is the equation for the decision hyper plane Fig. 3.19

The LHS of this equation gives the discriminate function $G(x)$. so that

$$w_1 x_1 + w_2 x_2 + \ldots\ldots w_n x_n + w_{n+1} = 0 \quad \text{...(3.3.6)}$$

or $$w^T x + w_{n+1} = 0$$

$$\Rightarrow \quad \begin{bmatrix} w \\ \hline w_{n+1} \end{bmatrix} T \begin{bmatrix} x \\ 1 \end{bmatrix} = 0 \quad \text{...(3.3.7)}$$

Here $$w = x_1 - x_2 \text{ and } w_{n+1} = \frac{1}{2}(\| x_2 \|^2 - \| x_1 \|^2) \quad \text{...(3.3.8)}$$

Squared distances can be computed from

$$\| x_0 - x_i \|^2 = x^T x - 2x_i^T x + x_i^T x_2 \text{ for } i = 1,2\ldots R \quad \text{...(3.3.9)}$$

As $x^T x$ is independent of R it is adequate to evaluate

$2x_0^T x + x_i^T x_i$ For $i = 1\ldots R$ and determine for which this term takes the largest of all values.

It may be noted that selection of the largest of $x_i^T x - 0.5 x_i^T x_i$ is same as the selection of smallest distance $\|x - x_2\|$.

The discriminate function becomes

$G_i(x) = w_i^T + wi_{n+1}$ for $i = 1\ldots R$

So we have $w_i = x_i$ and $w_{in+2} = -\frac{1}{2} x_i^T$ for $i = 1...R$; ...(3.3.10)

The minimum distance classifiers can be considered as linear classifiers or linear machines. This method is also called correlation classification.

3.3.5.2 *Correlation Classification*

The decision surface S_{ij} or the contiguous decision region. x_i, x_j is given by a hyper plane having the equation.

$$G_i(x) - G_j(x) = 0$$

$$w_i^T x + w_{i_{n+1}} - w_i^T x - w_{i_{n+1}} = 0 \quad ...(3.3.11)$$

The augmented vector of pattern is normally $\begin{bmatrix} x \\ 1 \end{bmatrix}$ so that the compact form of equation becomes $G_i(y) = \boldsymbol{w}_i^T \boldsymbol{y}$. Where y_i is the augmented vector. A linear classifier is shown in Fig. 3.20.

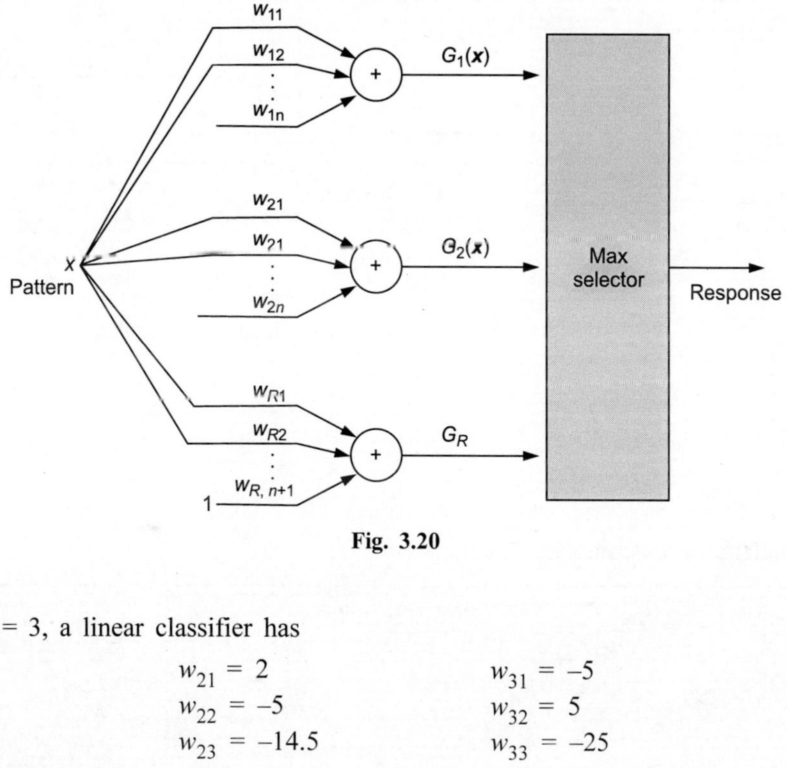

Fig. 3.20

Example:

For $n = 2$, $R = 3$, a linear classifier has

$w_{11} = 10$	$w_{21} = 2$	$w_{31} = -5$
$w_{12} = 2$	$w_{22} = -5$	$w_{32} = 5$
$w_{13} = 52$	$w_{23} = -14.5$	$w_{33} = -25$

3.3.6 Linearly Separable Patterns

Let a Pattern set X having subsets X_1, X_2, X_R be classified by a linear machine such that patterns can be classified clearly as X_i with $i = 1, 2...R$. in such a case the pattern sets are called Linearly Separable. So if R linear functions of $\boldsymbol{x}$ exist such that

$$G_i(\boldsymbol{x}) > G_j(\boldsymbol{x}) \text{ for } \boldsymbol{x} \in X_i \qquad i = 1, 2....R \; j = 1, 2...R \quad i \neq j$$

Then the pattern is said to be linearly separable. (XOR $(x_1\, x_2....x_n)$.

$= x_1 \oplus x_2 \oplus .....x_n$ is known as a linearly non-separable pattern set). Fig. 3.21 shows cases of 2-D and 3-D non-separability.

Using $R = 3$ and and $w_{i_{n+1}} = -\frac{1}{2} x_i^T x_i \; (n = 2)$

The weight vectors can be obtained as

$$w_1 = \begin{bmatrix} 10 \\ 2 \\ -52 \end{bmatrix} \quad w_2 = \begin{bmatrix} 2 \\ -5 \\ -14.5 \end{bmatrix} \quad w_3 = \begin{bmatrix} -5 \\ 5 \\ 25 \end{bmatrix}$$

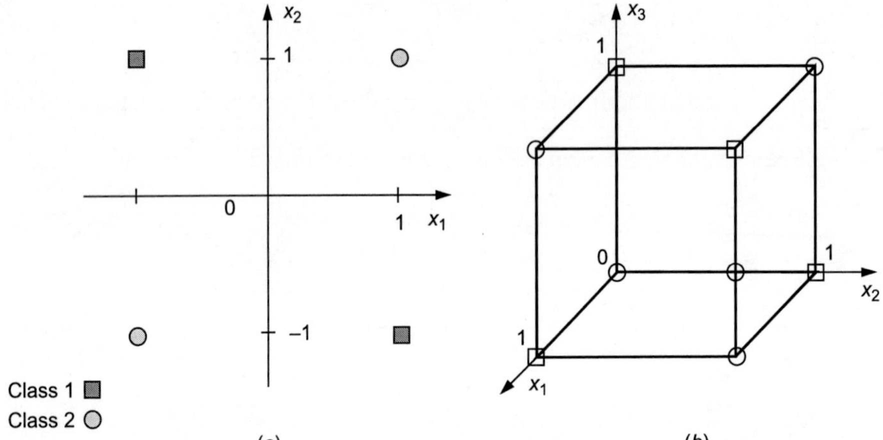

(a) (b)

Fig. 3.21

$$\left[W_I^3 = -\frac{1}{2}(10*10+2*2), W_{23} = -\frac{1}{2}(2*2-5), W_{33} = -\frac{1}{2}(-5*-5+5*5) \right]$$

$$\Rightarrow \quad G_1(x) = 10x_1 + 2x_2 - 52$$

$$G_2(x) = 2x_1 - 5x_2 - 14.5$$

$$G_3(x) = 5x_1 + 5x_2 - 25$$

We can tabulate the discriminate functions as:

Discriminate	Class 1 $[10\ 2]^T$	Class 2 $[2\ -5]^T$	Class 3 $[-5\ 5]^T$
$G_1(x)$	$(10x_1 + 2x_2 - 52)$ $= 10 * 10 + 2 * 2 - 52$	$(10x_1 + 2x_2 - 52)$ $= 10 * 2 - 2 * 5 - 52$	$(10x_1 + 2x_2 - 52)$ $= 10 * (-5) + -5 * 5 - 52$
	52	– 42	– 127
$G_2(x)$	$(2x_1 - 5x_2 - 14.5)$ $= 2 * 10 - 5 * 2 - 14.5$	$(2x_1 - 5x_2 - 14.5)$ $= 2 * 2 - 5 * (-5) - 14.5$	$(2x_1 - 5x_2 - 14.5)$ $= 2 * (-5) -5 * 5 - 14.5$
	– 4.5	14.5	– 49.5
$G_3(x)$	$(-5x_1 + 5x_2 - 25)$ $= -5 * 10 + 5 * 2 - 25$	$(-5x_1 + 5x_2 - 25)$ $= -5 * 2 + 5 * (-5) - 25$	$(-5x_1 + 5x_2 - 25)$ $= -5 * (-5) + 5 * 5 - 25$
	–65	60	25

It may be seen that diagonal terms have the highest value in comparison with the other columns.

The threshold values are $w_{1,3} = 52$ $\quad w_{2,3} = 14.5$ $\quad$ and $\quad w_{3,3} = 25$

The decision lines are calculated as:

$$\begin{cases} S_{12} : & G_1(x) - G_2(x) = 8x_1 + 7x_2 - 37.5 = 0 \\ S_{13} : & G_1(x) - G_3(x) = -15x_1 + 3x_2 + 27 = 0 \\ S_{23} : & G_2(x) - G_3(x) = -7x_1 + 10x_2 + 10x_2 - 10.5 = 0 \end{cases}$$

3.4 MULTI CATEGORY AND PERCEPTRON

The basic assumption here is that classes are linearly pair wise separable so that

$$G_i(\boldsymbol{x}) > G_j(\boldsymbol{x}) \quad \text{for} \quad i, j = 1, 2....R, \qquad i \neq j$$

Let us define an augmented weight vector w_p as

$$\boldsymbol{w}_p = [w_{p1}, w_{p2}...w_{pn+1}]^T \qquad ...(3.4.1)$$

An augmented pattern y of classic is presented to the maximum selector based classifier shown in Fig. 3.22. If $w_i^T y$ is longer than any of the other R–1 discrimant functions, no adjustment of weight vector is necessary as the classification is correct. The changed weights are:

$$\boldsymbol{w}_1' = \boldsymbol{w}_1 \quad \boldsymbol{w}_2' = w_2 ... \boldsymbol{w}_R' = \boldsymbol{w}_R \qquad ...(3.4.2)$$

If for some q value, $\boldsymbol{w}_i^T y \leq \boldsymbol{w}_q^T y$, then the updated vectors are

$$\boldsymbol{w}_i' = \boldsymbol{w}_i + cy \quad \boldsymbol{w}_q' = \boldsymbol{w}_q - cy \quad \boldsymbol{w}_K' = \boldsymbol{w}_k \text{ For } k = 1, 2...R \text{ where } k \neq i, q$$

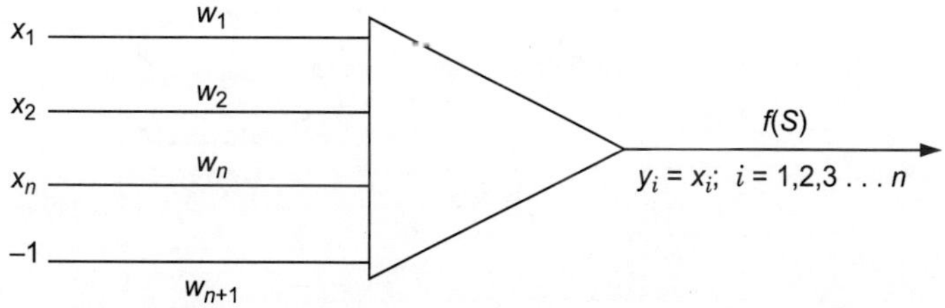

Fig. 3.22

In the full expression, the equations are

$$\boldsymbol{w}_{i_j}' = \boldsymbol{w}_{i_j} + cy_j \quad \boldsymbol{w}_{q_j}' = \boldsymbol{w}_{qj}' - cy_j, \quad \boldsymbol{w}_{K_j}' = \boldsymbol{w}_{K_j}$$

$$j = 1, 2...n+1, \quad k = 1, 2,...R \quad K \neq i, q, \ j = 1, 2...n+1$$

For the discrete case,

$$\boldsymbol{w}_i^{K+1} = \boldsymbol{w}_i^K + \frac{c^K}{2}(d_i^K - \text{output}_i^k)\, \boldsymbol{y}^k \text{ for } i, = 1, 2,...R; \qquad ...(3.4.3)$$

For R-category classifiers with local representation, the desired responses are

$$d_i = 1 \qquad d_j = -1 \quad \text{for } j = 1, 2......R \ j \neq i$$

For distributed representation, the above condition is not required as more than a single neuron is permitted to respond +1.

R-Category Discrete Perceptron Training Algorithm:

1. The Q training pairs are $\{x_1d_1, x_2d_2 ... x_qd_q\}$ first $c > 0$ is chosen
2. Weights are initialized at W at some random values.

$$q \leftarrow 1;\ k \leftarrow 1;\ E \leftarrow 0$$

3. $(\text{Output})_i \leftarrow \text{Sgn}\ (\mathbf{w}_i^T \mathbf{y})\ i = 1, 2....R.$

$$y \leftarrow y\ d \leftarrow d_q$$

4. $\mathbf{w}_i \leftarrow \mathbf{w}_i + \frac{1}{2} C(d_i - \text{output}_i)\ \mathbf{y}$

$$i = 1, 2....R$$

5. $E \leftarrow \frac{1}{2}(d_i - \text{output}_i)^2 + E$

$$i = 1, 2....R$$

6. If $q < Q$, $q \leftarrow q + 1$ $k \leftarrow k + 1$ go to 3 else go to 7.
7. Training Cycle is completed with $E = 0$.

Fig. 3.23 shows R-category linear classifier using R discrete perception.

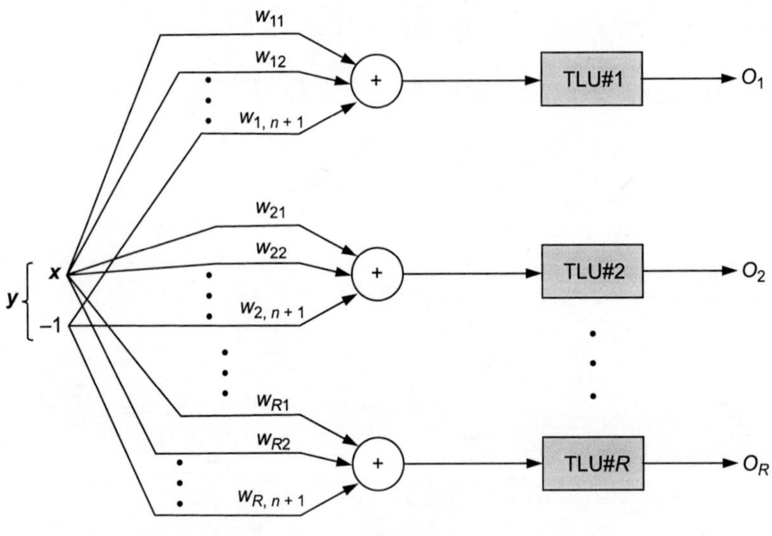

Fig. 3.23

3.5 TRAINING ALGORITHMS

3.5.1 Non-Parametric Training Concept

So far, the coefficients of linear discriminate functions (weights) have been determined based on prior information about sets of patterns and their class membership positions.

In non-parametric training the major steps are

(*a*) The sample pattern vectors $\boldsymbol{x}_1$, $\boldsymbol{x}_2$, $\boldsymbol{x}_p$ (training sequence) are presented to the machine along with the correct response which is provided by the teacher.
(*b*) The response specifies the classification information of each input vector.
(*c*) The parameters of the classifier are modified by iterations through supervised learning.
(*d*) The network learns from experience by finding the difference between responses (target and actual). The error value is used for adjusting the classifier. 4 TLU based perceptron serves this concept as shown in Fig. 3.24.

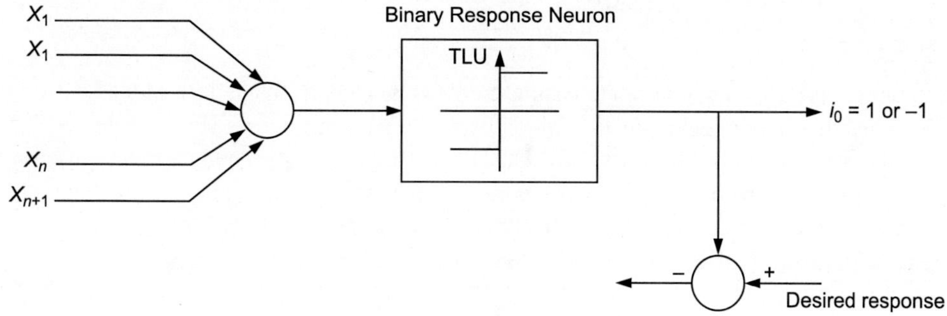

Fig. 3.24

Weight Solution Area (WSA)

In the first case, i.e. Fig. 3.25(*a*), it is assumed that initial weights $\boldsymbol{w}^1$ are for input $\boldsymbol{y}$, such that $\boldsymbol{w}^{1T}\boldsymbol{y}_1 < 0$. The weights are located in negative half plane. The pattern is misclassified. We have to increase the discriminant function $w^{1T}y$ by adjusting the weight vector. The best direction is in steepest increase corresponding to the gradient of discriminate function ($\boldsymbol{y}_1$). The adjusted weights become $\boldsymbol{w}^1 + qy_1$ where q is the positive valued correction constant used for controlling the degree of adjustment. Iterations (repeated weight adjustment) bring the weighted into WSA. The geometrical conditions of the augmented weight space are shown in Fig. 3.25.

In the second case, $\boldsymbol{w}^{1T}\boldsymbol{y}_1 > 0$, y_1 is of class II and the discriminant function has to be adjusted on the steepest decrease leading to new weight to be $\boldsymbol{w}^1 - qy_1$.

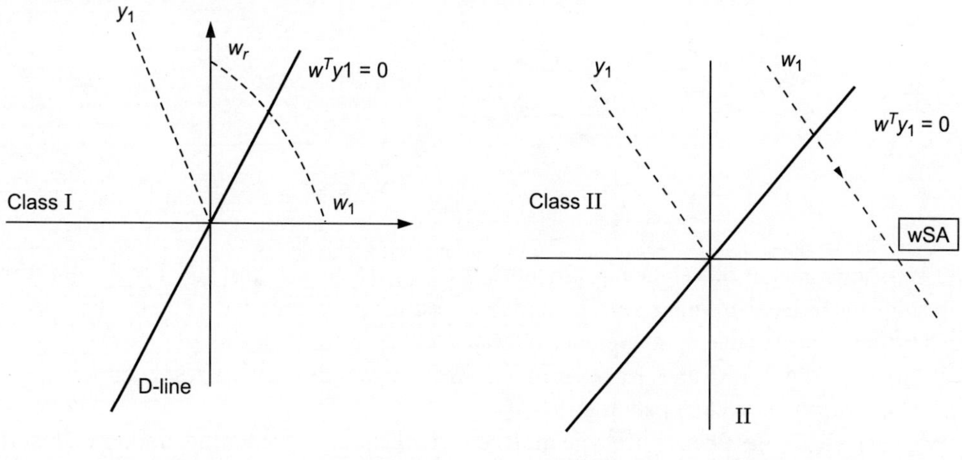

Fig. 3.25 *Weight Solution Area (WSA)*

3.5.2 Training Steps

1. Training starts at $\boldsymbol{w}^1$. In the beginning first prototype pattern of class 1 is presented. For misclassification, the new weights are

$$\boldsymbol{w}^2 = \boldsymbol{w}^1 + q\boldsymbol{y}_1$$

2. When the second pattern is presented for w^2 under misclassification,

$$\boldsymbol{w}^3 = \boldsymbol{w}^2 + q\boldsymbol{y}_2$$

3. As the third pattern is of class2, for misclassification

$$\boldsymbol{w}^4 = \boldsymbol{w}^3 - c\boldsymbol{y}_3$$

So the input y is equivalent to (j) increasing the $G(\boldsymbol{y})$ by $q\|y^2\|$ if the pattern of class I is undetected and (ii) decrease the discriminant function $q\|\boldsymbol{y}^2\|$ by if the pattern of class II is undetected. Change in weight is in the direction of the shortest path. The supervised training procedure can be summed up as:

Augmented weight vector = $\boldsymbol{w} \pm q\boldsymbol{y} + \rightarrow$ undetected pattern of class I

(For Correct classification, no adjustment is necessary) undetected path of class II

Weight adjustments of learning dichotomiser:

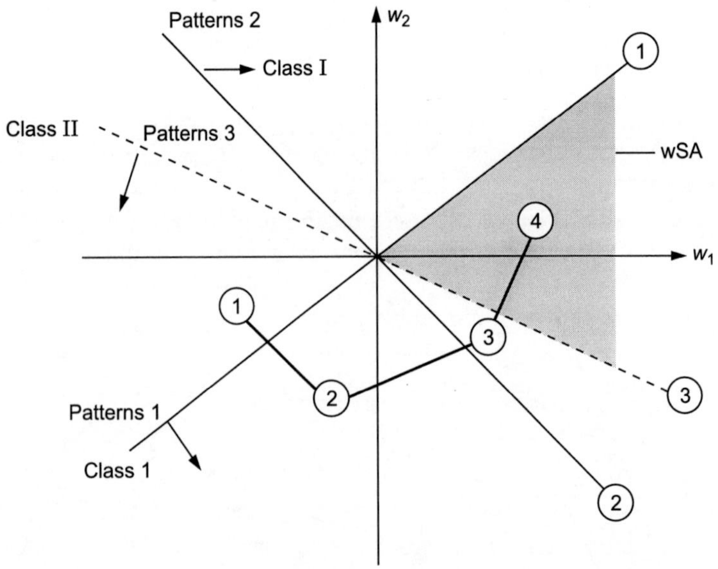

Fig. 3.26

Fig. 3.26 shows a series of weight adjustments in the augmented weight space. The weight corrections are shown for an ordered training set of inputs consisting of augmented vectors. The patterns are numbered as they appear at the input. Weights in the shaded region provide the solution for the augmented weight vector. The lines 1, 2 and 3 are fixed for variable weights which result during the training process. (as their normal vectors/patterns are fixed).

During 1st step, the first prototype pattern of class I is presented and misclassified so $\boldsymbol{w}^2 = \boldsymbol{w}^1 + c\boldsymbol{y}_1$. The second pattern is presented for the current weight vector w^2 and is misclassified $\boldsymbol{w}^3 = \boldsymbol{w}^2 + c\boldsymbol{y}_2$. Third pattern belongs to class II. This misclassification results in $\boldsymbol{w}^4 = \boldsymbol{w}^3 - c\boldsymbol{y}_3$.

Example: The classifier is to provide the classification of four patterns $\boldsymbol{x}$.

$x_1 = 1 \qquad x_3 = 3 \qquad d_1 = d_3 = 1 \qquad$ class I

$x_2 = -0.5 \qquad x_4 = -2.5 \qquad d_2 = d_4 = -1 \qquad$ class II

$$y_1 = \begin{bmatrix} 1 \\ 1 \end{bmatrix} \quad y_2 = \begin{bmatrix} -0.5 \\ 1 \end{bmatrix} \quad y_3 = \begin{bmatrix} 3 \\ 1 \end{bmatrix} \quad y_4 = \begin{bmatrix} -2.5 \\ 1 \end{bmatrix} \quad \text{Augmented vectors}$$

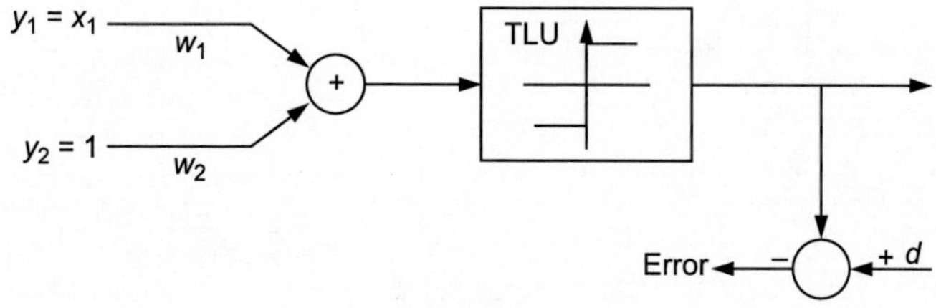

Fig. 3.27

Decision lines $\boldsymbol{w}^T\boldsymbol{y}_i = 0$ for $i = 1, 2, 3, 4$ are sketched in the augmented weight space along with their corresponding normal vectors ($\boldsymbol{y}_i = 1, 2, 3, 4$).

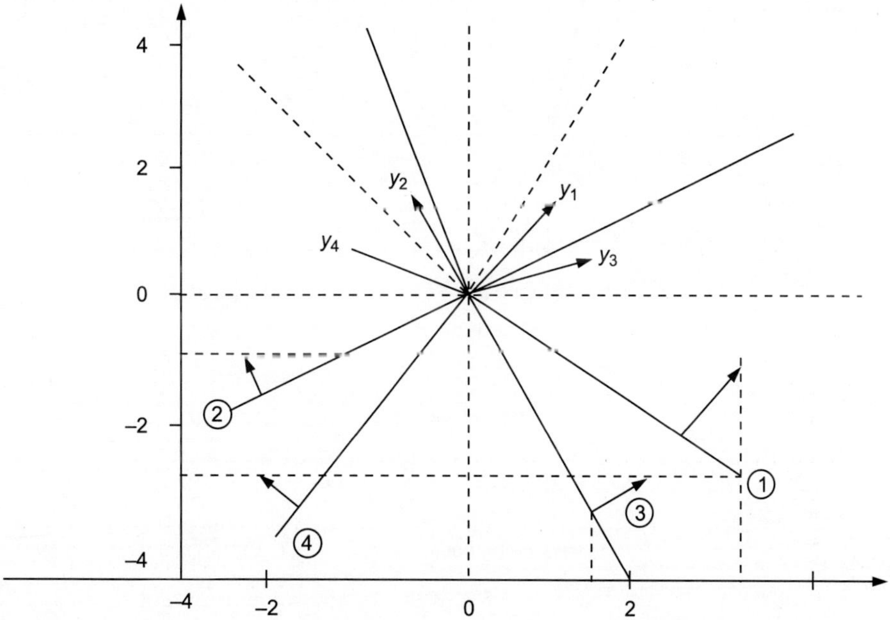

Fig. 3.28

Taking $c = 1$ and $\boldsymbol{w}^1 = [-2.5 \quad 1.75]^T$

Step 1: y_1 is the input

$$\text{Output}_1 = \text{Sgn } [(-2.5 \ 1.75)\begin{pmatrix} 1 \\ 1 \end{pmatrix}] = -1$$

$$d_1 - \text{output1} = 2$$

$$\boldsymbol{w}^2 = \boldsymbol{w}^1 + \boldsymbol{y}^1 = \begin{bmatrix} -1.5 \\ 2.75 \end{bmatrix}$$

Step 2: y_2 is the input

$$\text{Output}_2 = \text{Sgn}\left[(-1.5 \quad 2.75)\begin{pmatrix}-0.5\\1\end{pmatrix}\right] = 1$$

$$d_2 - \text{output}_2 = -2$$

$$\boldsymbol{w}^3 = \boldsymbol{w}^2 - \boldsymbol{y}^2 = \begin{bmatrix}-1\\1.75\end{bmatrix}$$

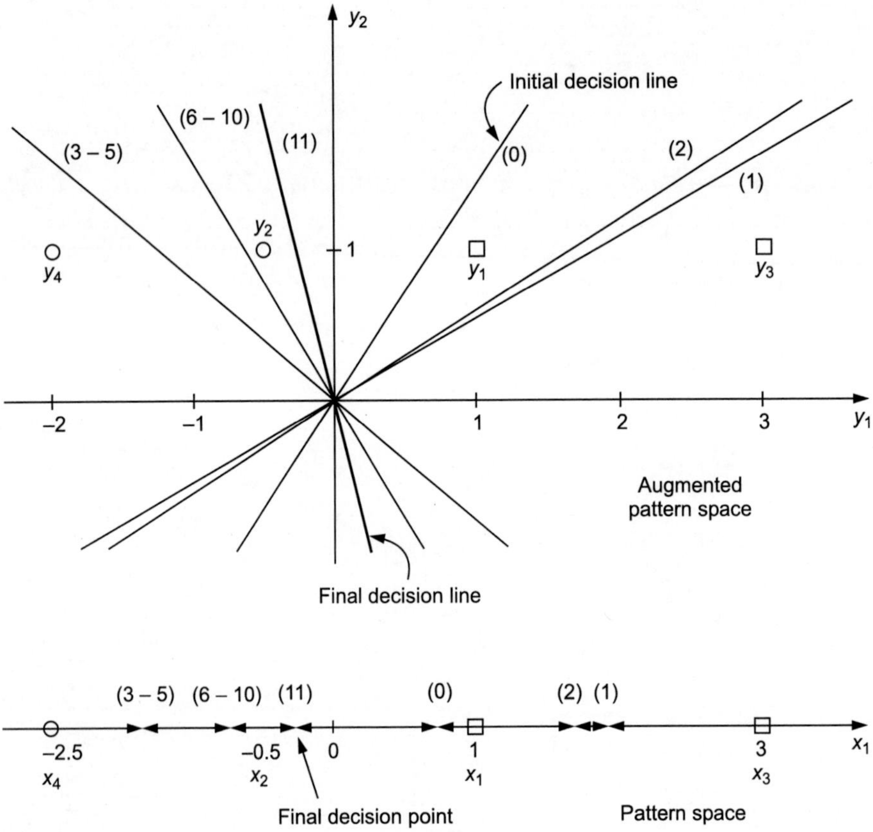

Fig. 3.29

Step 3: y_3 is the input

$$\text{Output}_3 = \text{Sgn}\left[(-1 \; 1.75)\begin{pmatrix}3\\1\end{pmatrix}\right] = -1$$

$$d_3 - \text{output}_3 = 2$$

$$\boldsymbol{w}^4 = \boldsymbol{w}^3 + \boldsymbol{y}^3 = \begin{bmatrix}2\\2.75\end{bmatrix}$$

As there is no proof of correct classification of w^4, we can take $\boldsymbol{y}^4 = \boldsymbol{y}_1$, $\boldsymbol{y}^5 = \boldsymbol{y}_2$, $\boldsymbol{w}^6 = \boldsymbol{w}^5 = \boldsymbol{w}^4$.

The later steps yield $w^7 = [2.5 \quad 1.75]^T$, $w^{10} = w^9 = w^8 = w^7$

The steps 6 and 10 have had weight adjustments. w^{11} ends up in solution area $\mathbf{w}_1 = 3$, $w_2 = 0.75$. Here $\mathbf{w}^{11}$ represents the solution weight vector and provides correct classification. In the augmented pattern space (y_1, y_2) the decision surfaces are straight lines. In the pattern space the decision surfaces are nearly point in x_1 axis.

3.5.3 Algorithm

The distance of point $\mathbf{w}^1$ from the plane $\mathbf{w}^T\mathbf{y} = 0$ in $(n + 1)$ dimensional Eucledian space is given by the $\pm\dfrac{\mathbf{w}^{1T}\mathbf{y}}{\|\mathbf{y}\|}$. The correction constant is to be selected so that $\mathbf{w}^2$ dislocates on the plane $\mathbf{w}^{1T}\mathbf{y} = 0$

$$\Rightarrow \qquad \mathbf{w}^{2T}\mathbf{y} = 0 \quad \text{or} \quad (\mathbf{w}^1 + c\mathbf{y})^T\mathbf{y} = 0$$

The length of the weight adjustment vector $(c\mathbf{y})$ is given by

$$\|c\mathbf{y}\| = \frac{|\mathbf{w}^{1T}\mathbf{y}|}{\mathbf{y}^T\mathbf{y}}\|\mathbf{y}\| = \frac{w^{1T}}{\|y\|} \quad \text{as} \quad \mathbf{y}^Ty = \|\mathbf{y}\|^2$$

So we can say that the distance of the point $\mathbf{w}^1$ to the decision plane is equal to the length of the weight increment vector. The correction constant depends on the current state of training program.

Depending on the length of the weight correction vector $\mathbf{w}^2 - \mathbf{w}^1$, a fractional coefficient can be introduced.

$$C = b\frac{|\mathbf{w}^{1T}\mathbf{y}|}{\mathbf{y}^T\mathbf{y}}$$

Fractional rule applies where $0 < b < 1$. Absolute correction rule is applicable when $1 < b < 2$. For the fixed correction rule with c = const, correction of weights is always the same fixed portion of current training vector.

In summary for classification into x_1 and x_2 the training task for the classifier is to have

$$\mathbf{w}^T\mathbf{y} > 0 \text{ for each } x \text{ of } x_1$$

$$\mathbf{w}^T\mathbf{y} > 0 \text{ for each } x \text{ of } x_2$$

At the completion of training with fixed training rule no further correction becomes necessary as no misclassification takes place.

Steps for Single Discrete Perceptron Training Algorithm (SDPTA)

1. The training pairs are $\{x_1d_1, x_2d_2 \ldots\ldots x_qd_q\}$, first $c > 0$ is chosen.
2. Weights are initialized at small random values.
3. Input is presented and output is computed for Sgn $(\mathbf{w}^T\mathbf{y})$. (k is the training step, q the step counter).
4. Weights are updated

$$\mathbf{w} \leftarrow \mathbf{w} + \frac{1}{2}C\,(d - \text{output})\,\mathbf{y}$$

(y is the augmented vector)

5. Cycle error is computed

$$E \leftarrow \frac{1}{2}(d - \text{output})^2 + E$$

6. If $q < Q$, $q \leftarrow q + 1$, $K \leftarrow K + 1 \rightarrow$ Go to step 3 else go to step 7
7. $E = 0$ Training cycle is completed.

Example: For non-parametric training of Discrete Perceptron the trained classifier is required to provide classification of four patterns x with known class membership. We have to find weights after 2 steps.

$x_1 = 1$	$x_3 = 2$	$d_1 = d_3 = 1$	class I
$x_2 = -1.5$	$x_4 = -2.0$	$d_2 = d_4 = -1$	class II

The single discrete perceptron with unknown weights w_1 and w_2 is to be trained using augmented input vectors.

$$y_1 = \begin{bmatrix} 1 \\ 1 \end{bmatrix} \qquad y_2 = \begin{bmatrix} -1.5 \\ 1 \end{bmatrix} \qquad y_3 = \begin{bmatrix} 2 \\ 1 \end{bmatrix} \qquad y_4 = \begin{bmatrix} -2.0 \\ 1 \end{bmatrix}$$

Let the initial weights be $w^1 = [-2 \quad 1.5)]^T$ (the constant is taken as 1)

$$\Delta \mathbf{w}^K = \frac{1}{2}[d_k - \text{Sgn}\,(\mathbf{w}^{KT}\, y_K)]\mathbf{y}^K$$

Step 1: The pattern $\mathbf{y}_1$ is the input

$$\text{Output}_1 = \text{Sgn}\left([-2 \ 1.5]\begin{bmatrix} 1 \\ 1 \end{bmatrix}\right) = \text{Sgn}\,[(-2) \times 1 + 1.5 \times 1]$$

$$= \text{Sgn}\,[-0.5] = -1$$

$$d_1 - \text{output}_1 = 2$$

$$\mathbf{w}^2 = \mathbf{w}^1 + \mathbf{y}^1 = \begin{bmatrix} -2 \\ 1.5 \end{bmatrix} + \begin{bmatrix} 1 \\ 1 \end{bmatrix} = \begin{bmatrix} -1 \\ 2.5 \end{bmatrix}$$

Step 2:

$$\text{Output}_2 = \text{Sgn}\left([-1 \quad 2.5]\begin{bmatrix} -1 \\ 1 \end{bmatrix}\right) = \text{Sgn}\,[-1 \times (-1) + 2.5 \times 1]$$

$$= \text{Sgn}\,[3.5] = 1$$

$$d_2 - \text{output}_2 = 2$$

$$\mathbf{w}^3 = \mathbf{w}^2 - \mathbf{y}^2 = \begin{bmatrix} -1 \\ 2.5 \end{bmatrix} + \begin{bmatrix} -1.5 \\ 1 \end{bmatrix} = \begin{bmatrix} 0.5 \\ 1.5 \end{bmatrix} \qquad \text{where } (\mathbf{y}^2 = \mathbf{y}_2)$$

Example: For a three class classifier based on discrete perceptions

$\boldsymbol{x}_1 = \begin{bmatrix} 10 \\ 2 \end{bmatrix}$ $\boldsymbol{x}_2 = \begin{bmatrix} 2 \\ -5 \end{bmatrix}$ $\boldsymbol{x}_3 = \begin{bmatrix} -4 \\ 4 \end{bmatrix}$. We have get the discriminant functions and analyse the classifier.

From the relation $\boldsymbol{w}_{i,\ n+1} = -\frac{1}{2}\boldsymbol{x}_i^T \boldsymbol{x}_i = -\frac{1}{2}[10 \quad 2]\begin{bmatrix} 10 \\ 2 \end{bmatrix} = -\frac{1}{2}[10\times10+2\times2] = -\frac{1}{4}104 = -52$

For thesecond case $-\frac{1}{2}\boldsymbol{x}_i^T \boldsymbol{x}_i = \frac{-1}{2}[4+25] = -14.5$

For the third case $-\frac{1}{2}x_i^T x_i = \frac{-1}{2}[-4\times4+4\times4] = -16$

So far $R = 3$, the weight vectors are augmented as

$$\boldsymbol{w}_1 = \begin{bmatrix} 10 \\ 2 \\ -52 \end{bmatrix} \quad \boldsymbol{w}_2 = \begin{bmatrix} 2 \\ -5 \\ -14.5 \end{bmatrix} \quad \boldsymbol{w}_3 = \begin{bmatrix} -4 \\ 4 \\ -16 \end{bmatrix}$$

The corresponding discriminant functions are:

$$G_1(x) = 10x_1 + 2x_2 - 52$$
$$G_2(x) = 2x_1 - 5x_2 - 14.5$$
$$G_3(x) = -4x_1 + 4x_2 - 16$$

In a tabular form, we can express the above as:

Discriminant	*Class I* $[10 \quad 2]^T$	*Class II* $[2 \quad -5]^T$	*Class III* $[-4 \quad 4]^T$
$G_1(x)$	$10x_1 + 2x_2 = -52$ for $x_1 = 10 x_2 = 2$ (52)	$10x_1 + 2x_2 - 52$ for $x_1 = 2,\ x_2 = -5$ (−42)	$10x_1 + 2x_2 - 52$ for $x_1 = -4\ x_2 = 4$ (−84)
$G_2(x)$	$2x_1 - 5x_2 - 14.5$ for $x_1 = 10 x_2 = 2$ (−4.5)	$2x_1 - 5x_2 - 14.5$ for $x_1 = 2,\ x_2 = -5$ (14.5)	$2x_1 - 5x_2 - 14.5$ for $x_1 = -4\ x_2 = 4$ (−42.5)
$G_3(x)$	$-4x_1 + 4x_2 - 16$ for $x_1 = -10\ x_2 = 2$ (−48)	$-4x_1 + 4x_2 - 16$ for $x_1 = 2\ x_2 = -5$ (44)	$-4x_1 + 4x_2 + 6$ for $x_1 = -4\ x_2 = 4$ (16)

The responses on the diagonal are the largest in each column choosing additional threshold. $T_1 = -2$ lowers the weight w_{13} from 52 to 50. By this TLU ≠ 1 responds with +1 for class I input. Similarly if T_2 and T_3 and −2 TLU ≠ 2 and TLU ≠ 3 respond appropriately, the design of three perceptron classifier is shown in Fig. 3.30.

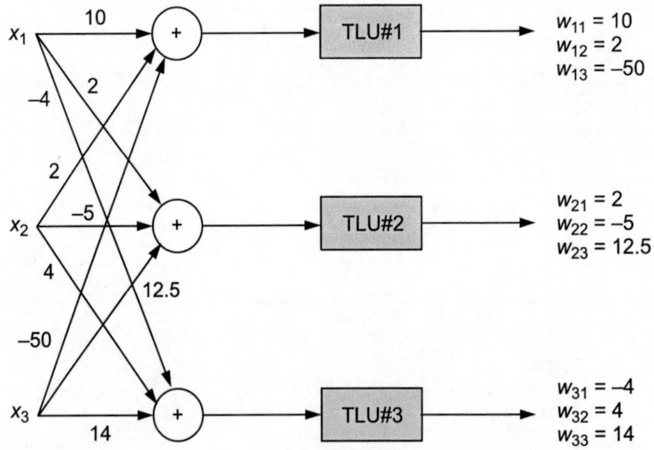

Fig. 3.30 *Three Perceptron Classifier from Maximum Selector.*

Example: Using $w_i^{K+1} = w_i^k + \frac{c^k}{2}(d_i^k - O_i^k)\boldsymbol{y}^k$ for 1 = 1, 2 ... *R*. Training of three perceptrons can be done. (Here O_i is output *i*).

Let the randomly chosen initial vectors be

$$w_1^1 = \begin{bmatrix} 1 \\ -2 \\ 0 \end{bmatrix} \qquad w_2^1 = \begin{bmatrix} 0 \\ -1 \\ 1 \end{bmatrix} \qquad w_3^1 = \begin{bmatrix} 1 \\ 3 \\ -1 \end{bmatrix}$$

Step 1: $\boldsymbol{y}_1$ is the input

$$\text{Sgn}\left[[1 \quad -2 \quad 0]\begin{bmatrix} 10 \\ 2 \\ -1 \end{bmatrix}\right] = 1$$

$$\text{Sgn}\left[[0 \quad -1 \quad 1]\begin{bmatrix} 10 \\ 2 \\ -1 \end{bmatrix}\right] = -1$$

$$\text{Sgn}\left[[1 \quad 3 \quad -1]\begin{bmatrix} 10 \\ 2 \\ -1 \end{bmatrix}\right] = 1$$

It is expected that the first case is 1 for correct response. –1 in the second also is expected. The +1 in third is not acceptable

Hence the only incorrect response is for TLU3

$$\boldsymbol{w}_1^2 = \boldsymbol{w}_1^1, \quad \boldsymbol{w}_2^2 = \boldsymbol{w}_2^1, \quad \boldsymbol{w}_3^2 = \begin{bmatrix} 1 \\ 3 \\ -1 \end{bmatrix} - \begin{bmatrix} 10 \\ 2 \\ -1 \end{bmatrix} = \begin{bmatrix} -9 \\ 1 \\ 0 \end{bmatrix}$$

Step 2: The values are 1, 1 –1. The case of +1 in the first case is in-appropriate. Hence

$$\mathbf{w}_1^3 = \begin{bmatrix} 1 \\ -2 \\ 0 \end{bmatrix} - \begin{bmatrix} 2 \\ -5 \\ -1 \end{bmatrix} = \begin{bmatrix} -1 \\ 3 \\ 1 \end{bmatrix} \qquad \mathbf{w}_3^3 = \mathbf{w}_3^2$$

$$\mathbf{w}_2^3 = \mathbf{w}_2^2$$

Step 3: Sgn $(\mathbf{w}_1^{3T}\ y_3) = 1$ Sgn $(\mathbf{w}_2^{3T}\ y_3) = -1$

Sgn $(\mathbf{w}_3^{3T}\ y_3) = 1$. The first case is incorrect.

$$w_1^4 = \begin{bmatrix} 4 \\ -2 \\ 2 \end{bmatrix} \qquad w_2^4 = w_2^3, \qquad w_3^4 = w_3^3$$

The subsequent steps yield $w_1^5 = w_1^4$, $w_1^6 = \begin{bmatrix} 2 \\ 3 \\ 3 \end{bmatrix}$ $w_1^7 = \begin{bmatrix} 7 \\ -2 \\ 4 \end{bmatrix}$

$w_1^8 = w_1^7\ w_1^9 = \begin{bmatrix} 5 \\ 3 \\ 5 \end{bmatrix}$. As none of the last three pattern presentation requires adjustment of weights, the training procedure is terminated. Figure 3.31 shows the 3-Perceptron network as a result of the training.

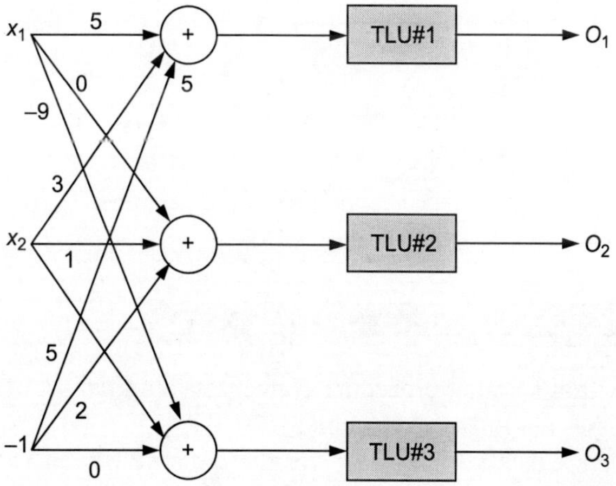

Fig. 3.31

Classification done by the above is

$O_1 = \text{Sgn}\ (5x_1 + 3x_2 - 5)$ $O_2 = \text{Sgn}\ (-x_2 - 2)$ $O_3 = \text{Sgn}\ (9x_1 + x_2)$

The resulting decision surfaces $(5x_1 + 3x_2 - 5) = 0$, $(-x_2 - 2) = 0$, $(9x_1 + x_2 = 0)$. Thus a triple dichotomization of the plane x_1, x_2 is achieved.

In contact to the minimum distance classifier (Fig. 3.31), this techniques produces several indecision regions. In such regions, no class membership of an input pattern can be uniquely determined based classifier.

3.6 PERCEPTRON NETWORKS

Linear classifier using Discrete Perceptrons (Fig. 3.31) is an example of discrete perceptron networks. The weight adjustment during n^{th} step is done by,

$$w_i^{k+1} = w_i^k + \frac{c^k}{2}(d_i^k - O_i^k)\, y^k \text{ for } 1 = 1, 2 \ldots R$$

There are two cases

(*a*) Distributed representation
(*b*) Local representation.

Based on Continuous Perceptrons, we can build R-category linear classifier as shown in Fig. 3.32.

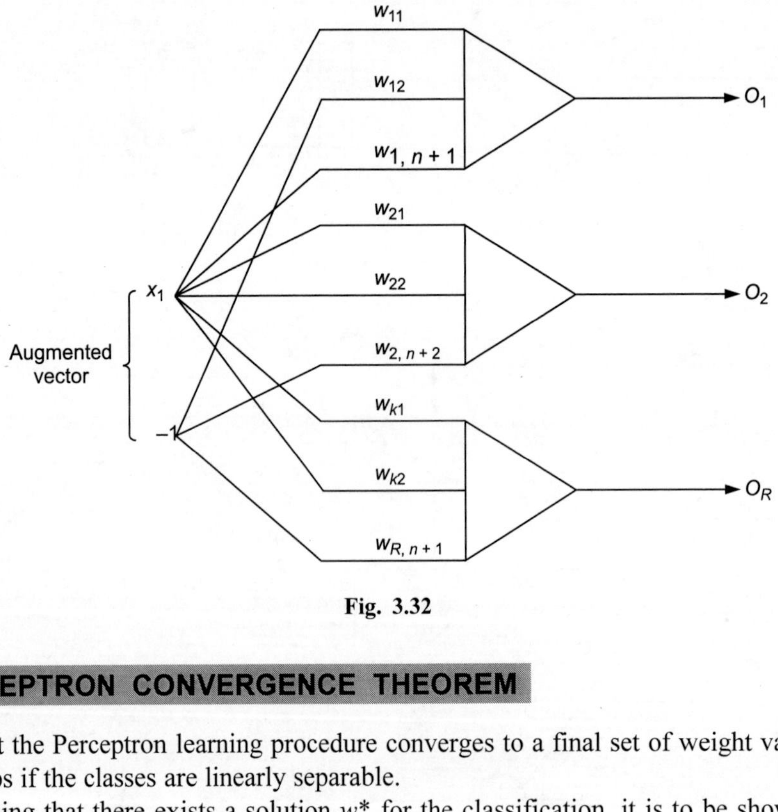

Fig. 3.32

3.7 PERCEPTRON CONVERGENCE THEOREM

This states that the Perceptron learning procedure converges to a final set of weight values in a finite number of steps if the classes are linearly separable.

Proof: Assuming that there exists a solution w^* for the classification, it is to be shown that w^* can be approached in a finite number of steps

$$w^{*T}x > \propto > 0 \quad \text{for each } x \in X_1$$

If we start from $w(0) = 0$, then

$$w(k) = \gamma \Sigma x(i)$$

$$[w(k+1) = w(k) + \gamma\, x(K) \text{ for } x \in X_1 \qquad \ldots(3.7.1)$$

Multiplying by w^{*T}, $\quad w^{*T}.w(k) = \gamma \Sigma\, w^{*T} x_i > \gamma \times k \times \propto \qquad \ldots(3.7.2)$

According to Cauchy Schwartz inequality

$$\|w^{*T}\|^2 \ \|w(k)\|^2 \geq \|w^{*T} \ w(k)\|^2 \qquad ...(3.7.3)$$

So

$$\| w(k) \|^2 > \frac{(\gamma \times k \times \infty)^2}{\left\| w^{*T} \right\|^2}$$

$$\|w(k + 1)\|^2 = (w(k) + \gamma x(k))^T \ (w(k) + \gamma \ x(k))$$

$$= \|w(k)\|^2 + \gamma^2 \ \| \ x(k)\|^2 + 2\gamma w^T(k) \ x(k) \qquad ...(3.7.4)$$

$$\leq \|w(k)\|^2 + \gamma^2 \ \| \ x(k)\|^2 \qquad ...(3.7.5)$$

From the classification condition

$$\|w(k)\|^2 \leq \gamma^2 \ \Sigma\|x_i\|^2 < \gamma^2 \ k\beta \qquad ...(3.7.6)$$

where $\beta = \max \ \|x_i\|^2$

So the optimum value of k is given by

$$\frac{K^2 \infty^2}{\| w^{*T} \|^2} = \beta \, k \text{ or } K = \frac{\beta}{\infty^2} \| w^* \|^2 \qquad ...(3.7.7)$$

As β is positive, the number of steps k will be finite.

3.8 LIMITATION OF PERCEPTRON MODEL

Convergence in Perceptron model takes place only if the pattern classes are linearly separable. The condition of linear separability requires that the convex hull of pattern sets of the classes are disjoint (a hull of a pattern set is the smallest convex set in the pattern space containing the pattern set). Basically convex set is a set of point in the N-dimensional space that a line joining any two points in the set lies fully in the region enclosed by the set. For the case of linear separation, the Perceptron Convergence theorem guarantees that the final set of weights will be attained in a finite number of sets. These weights, in turn, define a linear hyper plane separating the two classes. However, actually the number of linearly separable functions decreases rapidly as the dimension of input pattern space is increased. Following table shows the linearly separable functions for a two-class problem with binary input patterns.

Dimension of Input Data	*No. of Possible Function*	*No. of Linearly separable*
1	4	4
2	16	14
3	256	104
4	655536	1882
5	$\sim 4.3 \times 10^9$	49572
6	$\sim 1.8 \times 10^9$	15028134

The problem shown in Fig. 3.33(*a*) is that of linearly separable functions. There are two functions which are linearly inseparable as shown in Fig. 3.33(*b*). These inseparable problems do not lead to convergence of weights through the Perceptron learning techniques conforming that the problem cannot be represented by a single layer of Perceptron.

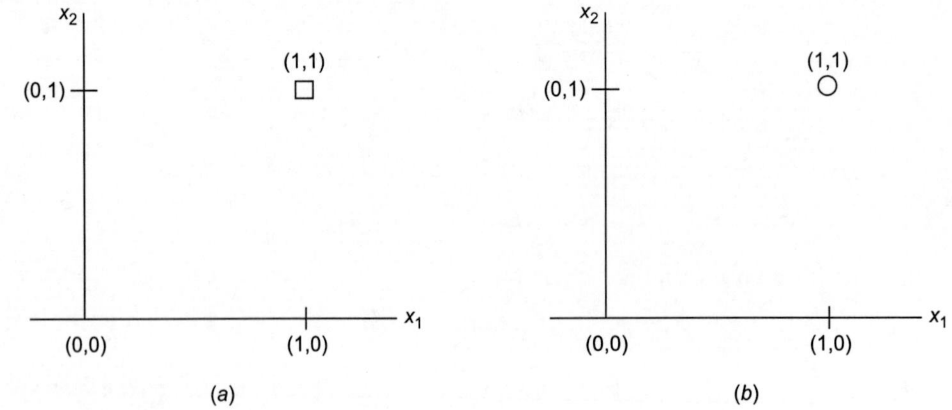

Fig. 3.33

Evidently the first type (*a*) corresponds to simple Logic function whereas (*b*) corresponds to EXOR Variety.

3.9 APPLICATIONS

(*a*) *Pattern Recognition and Image Processing:* Perceptron based networks have been widely used in pattern recognition and image processing based on classification techniques Fig. 3.33. Explains such possibilities.

There are many problems of linear separability which cannot be represented by a single Perceptron. Such "hard" problems have to be tackled by multi-layers Perceptron.

(*b*) *Optimisation/constraint satisfaction:* Examples are manufacturing, scheduling and shortest travel plan.

(*c*) *Fore-casting and Risk Management:* Examples are in meteorology (weather fore-casting) stock market, econometrics.

(*d*) *Control systems:* These include many commercial products launched by Japanese companies.

(*e*) Chemical plants and Robots.

SUMMARY

- This unit highlights the feed forward networks as well as the role of perception. Concepts classifiers.
- The discriminant have been lucidly developed so that complex topics like multi-category perceptron can be understood.
- This is almost the heart of the syllabus for neural networks and has to be appreciated and assimilated with care.

Short Questions

1. What are discriminant functions?
2. What is decision hyperplane?
3. What is dichotomiser?
4. How linear separability is linked to the perceptrons?
5. What is the non-parametric analysis?
6. What is the main use of discrete perceptron?
7. What is Cauchy-schwartz inequality?
8. What are the main limitations of perceptron model?
9. What is the background of Rosenblatt's perceptron?
10. What are the main applications of continuous perceptron?

Long Answer Questions

1. **What is perceptron? Explain briefly the architecture and algorithm of discrete percepton network.**

   Refer 3.2 and 3.6
2. **Discuss in detail about single layer and multi layer continuous perceptron networks.**

   Refer 3.4 and 3.4.1
3. **State and prove perceptron convergence theorem**

   Refer 3.7
4. **Explain Principle and algorithm of continuous perceptron.**

   Refer 3.6
5. **Write brief on multy category perceptron with diagram.**

   Define:

   (i) Pattern recognition and clossfier Refer 3.3.12

   (ii) Continuous discriminant function Refer 3.3.3

   (iii) Linear Classifier Refer 3.3.5.1
6. **Compare discrete and continuous perceptron network.**

   Refer 3.6
7. **Give applications of single longer feed forward network.**

   Refer 3.9

❑❑❑

4

MULTILAYER FEED FORWARD NEURAL NETWORKS

4.0 INTRODUCTION

In this unit, the credit assignment, generalized delta rules, Back Propagation and its merits, demerits are analyzed. Credit assignment problem is basically the job of assigning credit for the outcomes. The generalized delta rule is an extension of the delta rules for network consisting of several layers. Before getting into the mathematical analysis, various features of Back Propagation and error analysis have been taken up for discussion.

The error between the computed output and desired output needs to be reduced. This can be done by adjusting the weights. In case of hidden layers being present, the weight adjustment has to be carried out in two phases. In the first phase, the output errors have to be used to correct the weights between hidden layers and output layers. Next this correlation mechanism has to be "back propagated" to weights between input layers and hidden layers. This is the basis of Back Propagation.

Kolmogorov Theorem has been touched briefly. The merits, demerits, limitation and correction have been taken up for complete understanding of Back Propagation techniques. The basic technique of Back Propagation has been illustrated by using a popular example.

4.1 CREDIT ASSIGNMENT PROBLEM

According to Minsky, the credit assignment problem is the task of assigning credit or blame for overall outcomes to each of the internal decisions made by a learning machine and those contributed to the outcomes. It can be divided into

1. Temporal credit assignment problem which involves the instants of time when the actions that deserve credit were taken.
2. Structural credit assignment problem which involves assigning credit to the internal structures of actions generated by the system.

The first one is relevant when there are many actions taken by a learning machine that result in certain outcomes (We have to find out which of these actions are responsible for the outcomes).

The second one is relevant when a multi component learning machine is under consideration (we have to determine exactly which particular component should change its behaviour to the desired limit).

The credit assignment problem arises when error correction learning is applied to feed forward neural network. The hidden neurons and output neurons have to be considered.

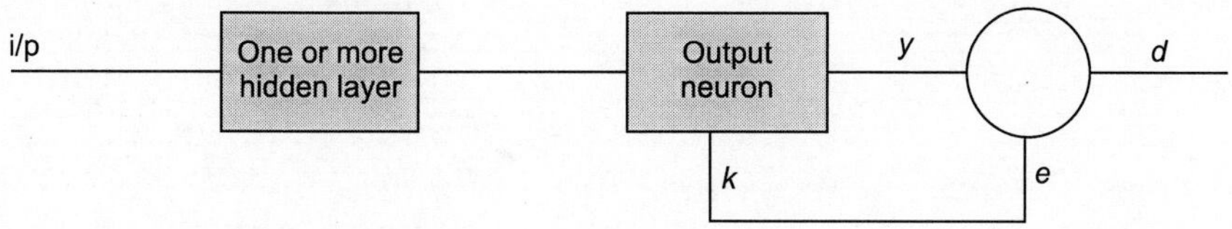

Fig. 4.1

For the output neuron, the desired output helps to assign credit. The case of the hidden layers has to be analyzed properly for the assignment of credit. Even though the hidden neurons are not directly accessible, they share responsibility for any error. Normally this problem is solved by the error signal.

4.2 GENERALISED DELTA RULE

Delta learning applied to Continuous Perceptrons can be generalised for a network consisting of several layers of feed forward units. This can also be used for back propagation. Prior to a more complex treatment of this rule it is worthwhile to have a simple view of the inside of a network.

There are three layers: an input layer, an output layer and a layer in between called a hidden layer. The neural structure of the brain is generally multilayer and has mostly a feed forward structure.

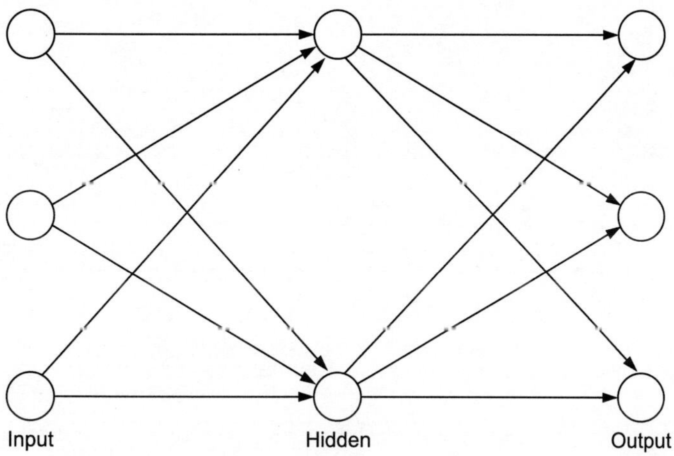

Fig. 4.2

Let the output of node i, u_i be sent to node j.

The input to mode j will be

$$x_j = \sum_i (w_{ij} u_j) \qquad \text{...(4.2.0)}$$

The output of node j will be $u_j = f_i \sum_i (w_{ij} u_{pi})$ ...(4.2.1)

where f_j is the activation function of layer j.

Let E_p be the measure of error defined by

$$E_p = \frac{1}{2} \sum_j (d_{pj} - u_{pj})^2 \qquad \text{...(4.2.2)}$$

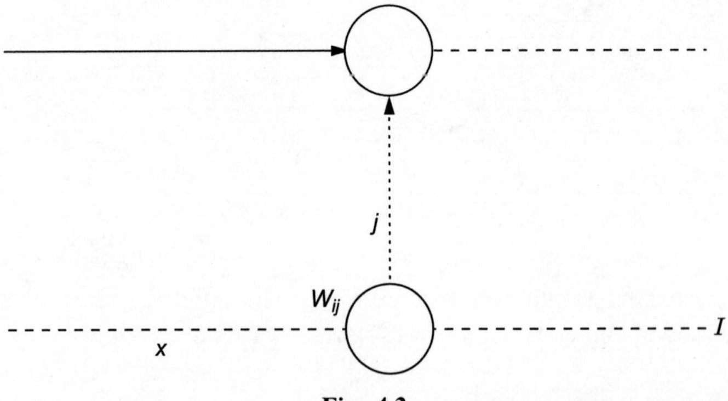

Fig. 4.3

where d_{pj} is target output from node j for pattern p and u_{pj} is the actual output.

$$\frac{\partial E_p}{\partial w_{ij}} = \frac{\partial E_p}{\partial x_{pj}} \cdot \frac{\partial x_{pj}}{\partial w_{ij}} \qquad ...(4.2.3)$$

Let the change in error as a function of the change inputs be

$$-\frac{\partial E_p}{\partial x_{pj}} = \delta_{pj} \qquad ...(4.2.4)$$

$$\frac{\partial x_{pj}}{\partial w_j} = \frac{\partial}{\partial w_{ij}}(\sum_k (w_{kj} u_{pk})) \quad \text{As} \quad x_j = \sum_i (w_{ij}, u_i)$$

$$= \sum_k (\frac{\partial w_{kj}}{\partial w_{ij}} . u_{pk}) \qquad ...(4.2.5)$$

As $\frac{\partial w_{kj}}{\partial w_{ij}} = 0$ except when $k = i$; $\frac{\partial x_{pj}}{\partial w_{ij}} = u_{pi}$;

$$-\frac{\partial E_p}{\partial w_{ij}} = \delta_{pj} . u_{pi} \qquad ...(4.2.6)$$

So we can write

From the above, it can be concluded that to decrease the value of *EP*, The weight changes $\Delta_p w_{ij}$ are to be proportional to $d_{pj} \cdot u_{pi}$

$$\Rightarrow \quad \Delta_p w_{ij} = \text{const } d_{pj} \cdot u_{pi} \qquad ...(4.2.7)$$

We may write $\delta_{pj} = -\frac{\partial E_p}{\partial x_{pj}} = -\frac{\partial E_p}{\partial u_{pj}} \cdot \frac{\partial u_{pj}}{\partial x_{pj}}$

We know that $u_{pj} = f_j(x_{pj}) \Rightarrow \frac{\partial u_{pj}}{\partial x_{pj}} = f_j'(x_{pj}) \qquad ...(4.2.8)$

Again $\frac{\partial E_p}{\partial pj} = -(d_p - u_{pi})$

The δ factor can be written as

$$\delta_{pj} = f_j'(x_{pj})\,(d_{pj} - u_{pj}) \qquad ...(4.2.9)$$

The above equation holds for the output units having d_{pj} as target.

If j is not an output unit,

$$d_{pj} = \sum_k \left(\frac{\partial E_p}{\partial x_{pk}} \cdot \frac{\partial x_{pk}}{\partial pj}\right)$$

$$= \sum_k \frac{\partial E_p}{\partial x_{pk}} \cdot \frac{\partial}{\partial y_{pj}} \left(\sum w_{ik} . u_{pi}\right) = -\sum_k \delta_{pk} w_{jk} \qquad ...(4.2.10)$$

We can write that $\delta_{pj} = f^1(x_{pj}) \sum_k (\delta_{pk} w_{jk})$

For a Sigmoid function, $f'(x_{pj}) = \text{const } y_{pj}\,(1 - u_{pj})$

4.3 DERIVATION OF BACK PROPAGATION

4.3.1 Feature of Back Propagation

Back Propagation learning is based on propagation of the errors backwards from the output layer to the hidden layers for determining the update for the weights leading to the units in the hidden layers .There is however, no feedback of the signal itself at any stage as the network concerned is feed forward neural network.

Back Propagation technique is synonymous with generalized delta rule as it is the same as delta learning at the output layers and is similar to delta learning for the hidden layers; the "generalized" factor is regarding the extension to the hidden layers.

Back Propagation of error is possible only if the output functions of the non-linear processing units are differentiable. In case of linear output functions, the advantage of a multilayer network for generating complex decision boundaries for a non-linearly separable or hard classification.

A multilayer feed forward network with only linear processing is equivalent to a linear associative network which has limitations of applications. Hard-limiting output functions (in multilayer Perceptron) can not be used for updating the weights. Generally, Sigmoidal non-linearity is used in Back Propagation.

[$f(x) = \dfrac{1}{1+\exp^{-\lambda x}} \Rightarrow \dfrac{df}{dx} = \lambda f(x)(1 - f(x)); f(x) = \tanh lx \Rightarrow \dfrac{df}{dx} = L(1 - f^2(x))$. [The former is called logistic function].

The Back Propagation method is based on gradient descent along the error surface. (Weight adjustment is taken as proportional to the negative gradient of the error with respect to the weight). We are only concerned with the instantaneous error between the desired and actual output values. As this is due to a given training pattern which is a sample function of a random process, the error can be assumed to be a random variable.

Thus the gradient descent method is basically stochastic type in which path to the minimum of the error surface will be zigzag. Actually the error surface itself becomes an approximation of the true error surface determined by the entire set of training patterns. In case of ADALINE, the true surface is quadratic; in general this quadratic quality is lost.

The error surface may contain several local minima beside the global minimum. These imply that due to stochastic approximation of the gradient decent used in the back propagation convergence is ruled out. (There is no convergence even the mean like the case of LMS algorithm). The four possibilities are:

The input output pair vectors are sample functions of random process. Linear units—The error surface $E(w)$ is the smooth hemispherical structure in the weight space with the gradient descent $\frac{\partial E}{\partial w_{uj}}$ converging to the optimal weight vector. Here we consider the ensemble average of the error.

For a finite set of input-output pairs the estimation of error is considered and in case of linear unit, there is an approximate convergence.

Instantaneous error, in case of linear units, has the converging gradient descent only in the mean.

For non-linear units, there is no proof of convergence even in the stochastic model.

Stopping of the iteration may be done by using some heuristic data based on values of gradient error in successive iterations and total number of iterations. Few of such possibilities are

1. The average gradient value over each training cycle is observed and this is below a preset threshold value for successive cycles, then the training process may be stopped.
2. A threshold for the average error may be taken and the average error in successive cycles is observed. Stopping is done by taking the threshold into consideration.
3. Both the average gradient and the average error may be used for stopping the process.

In a multilayer feed-forward network, synaptic weights of input layer-hidden layer and hidden layer-output layer may be represented by vectors V and W. These are analogous to the base of accumulation knowledge. These strengths are adjusted to reduce the gap between new input and the accumulated values. The purpose of learning is to make V and W more compatible with requirements of the current input.

By the method steepest descent the errors can be minimised. An error surface of the form $E = \Sigma_p E^p\ (V, W)$ can be shown as in Fig. 4.4.

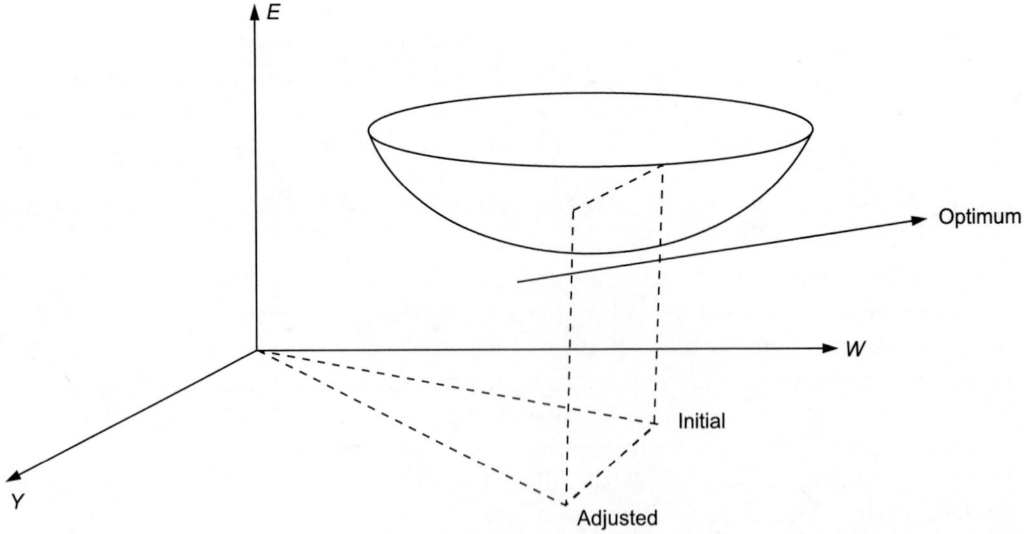

Fig. 4.4

In general, multilayer feed forward networks with non-linear activation functions have mean squared error surface above the multi-dimensional weight space (which is a smooth parabolic surface in single layer activation). Actually the error surface is more complex and consists of many local and global minima unlike the delta rule, finding minimum cannot be guaranteed in case of Back Propagation.

Let be initial weights be $W(0)$ and $V(0)$ and the training pairs may be represented by (IP, oP). By gradient descent, weights are moved in direction of quickest fall so that a different E location is reached on the error weight surface. This procedure of iteration is repeted until the error comes under acceptable limits or the number of training cycles is reached. Error surface can be described as summation of quadratic term which describes elliptical rather than circular contours. The gradient in variably does not point directly in path of minimum.

4.3.2 Derivation of Back Propagation

The gradient in the error space can be shown as,

$$\Delta = \frac{\partial E}{\partial V}\bar{u}_1 + \frac{\partial E}{\partial W}\bar{u}_2 \qquad ...(4.3.3)$$

Weight change $\Delta = V\bar{u}_1 + \frac{\partial E}{\partial W}\bar{u}_2$ can be expressed as –negative constant as multiplied by gradient.

Component wise, we may write,

$$\Delta V = -\eta\frac{\partial E}{\partial V} \qquad \Delta W = -\eta\frac{\partial E}{\partial W}$$

E_k for the kth neuron is given by, $E_k = \frac{1}{2}(d_k - O_k)^2$ ...(4.3.4)

Where dk is the target output and O_{OK} is the computed output of Wth output neuron.

as
$$I_{OK} = W_{1k}O_{H1} + W_{2k}O_{H2} + W_{mk}O_{IIm}$$

$$\frac{\partial E}{\partial W_{ik}} = \frac{\partial E_K}{\partial 0_{OK}}\frac{\partial O_{OK}}{\partial I_{OK}}\frac{\partial I_{OK}}{\partial W_{iK}} \qquad ...(4.3.5)$$

Where $\frac{\partial E_K}{\partial O_{OK}} = -(d_K - O_{OK})$ and $\frac{\partial O_{OK}}{\partial I_{OK}} = \frac{1}{\lambda O_{OK}(1-O_K)}, \frac{\partial I_{OK}}{\partial W_{iK}} = OH_2$

$$\Delta W_{ik} = -\eta\frac{\partial E_k}{\partial W_{ik}} = \eta[\lambda(T_k - O_k)\,O_{ok}(1-O_{ok})\,O_{Hi}] \qquad ...(4.3.6)$$

$$[\Delta W] = -\eta\{O\}_H \times \lambda < (d_k - O_k)\,O_{OK}(1-O_{OK}) = n\{O\}_H < l > \qquad ...(4.3.7)$$

For carrying out the change in hidden layer,

$$\frac{\partial E_k}{\partial V_{ij}} = \frac{\partial E_k}{\partial O_{Ok}}\frac{\partial O_{OK}}{\partial I_{OK}}\frac{\partial I_{OK}}{\partial O_{Hi}}\frac{\partial OH_i}{\partial IH_j}\frac{\partial IH_j}{\partial V_{ij}} \qquad ...(4.3.8)$$

Where $\frac{\partial I_{OK}}{\partial O_{Hi}} = W_{ik}$ and $\frac{\partial OH_i}{\partial HI_j} = \lambda(O_{Hi})(1-O_{Hi})$ and $\frac{\partial IH_j}{\partial V_{ij}} = O_{Ij}$

$$\frac{\partial E_k}{\partial O_{OK}}\frac{\partial O_{OK}}{\partial I_{OK}}\frac{\partial I_{OK}}{\partial O_{Hi}} = -W_{ik} \times l_k = -g_i \qquad ...(4.3.9)$$

If we define l_k^* as $l_k^* = -g_i\lambda(O_{Hi})(1-O_{Hi})$

$$\frac{\partial E_k}{\partial V_{ij}} = -l_i^* I_{ij}$$

or

$$[\Delta V] = n\{I\}_I \{l^*\}$$

$$l_i^* = -g_i\lambda(O_{Hi})(1-O_{Hi}) \qquad ...(4.3.10)$$

Finally,

$$[\Delta W] = \eta\{O\}_H < l > \text{ and } [\Delta V] = n\{I\}^I \{l^*\} \qquad ...(4.3.11)$$

4.3.3 Back Propagation Learning Algorithm

1. Neural networks work better if input and output lie between 0 to 1. So the input and output should be normalized.
2. Let $[\Delta V]^O = [\Delta W]^O = [O]$
3. For the training data, one set of inputs and outputs must be presented. The pattern to the input layer $\{I\}_I$ become the input to the input layer because of linearity $\{O\}_I = \{I\}$
4. $\{I\}_H$ = Input to the hidden layer = $[V]^T \{O\}^I$

   the function at hidden layer are sigmoid so $\{O\}_H = \begin{Bmatrix} | \\ | \\ \dfrac{1}{1+e^{-I_{H_i}}} \\ | \\ | \end{Bmatrix}$

   Input to the output layer is found from $\{I\} = \{W\}^T \{O\}_H$

$$\{O\}_H = \begin{Bmatrix} | \\ | \\ \dfrac{1}{1+e^{-I_{Oj}}} \\ | \\ | \end{Bmatrix}$$

5. Error $E^P = \dfrac{\sqrt{\Sigma d_k - O_{oj}}}{N_o}$
6. Determine $\{e\} = \{(d_k - O_k)O_{ok}(1-O_{ok})\}$
7. Determine $\{e^*\}$.
8. These correction matrices give the weight updating

Example:

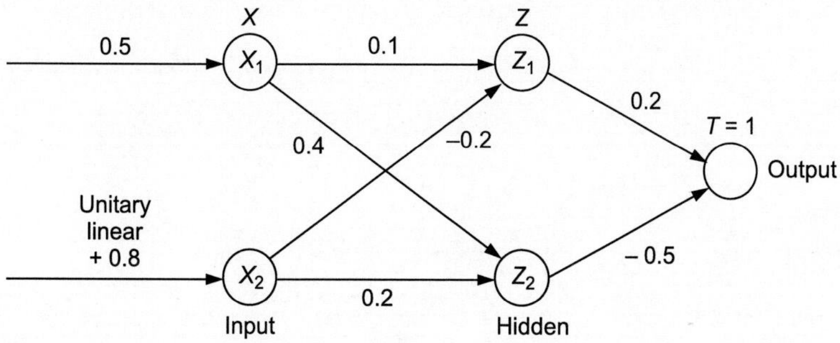

Fig. 4.5

Training set		*Output O*
I_1	I_2	o
0.5	–0.8	0.1

Step 1:

$$\{O\}_I = \{I\}_I = \begin{bmatrix} 0.5 \\ +0.8 \end{bmatrix}$$

Step 2: Weights are initialized

$$[V]^0 = \begin{bmatrix} 0.1 & 0.4 \\ -0.2 & 0.2 \end{bmatrix}$$

$$[W]^0 = \begin{Bmatrix} 0.2 \\ -0.5 \end{Bmatrix}$$

Step 3: We have to determine the input to hidden layer from the input to hidden layer and V matrix

$$\{I\}_H = [V]^T \{O\}_T$$

$$= \begin{bmatrix} 0.1 & -0.2 \\ +0.4 & 0.2 \end{bmatrix} \begin{Bmatrix} 0.5 \\ -0.8 \end{Bmatrix} = \begin{Bmatrix} -0.11 \\ +0.36 \end{Bmatrix}$$

Step 4: For sigmoid function the output of hidden layer is given by

$$\{O\}_H = \begin{Bmatrix} \dfrac{1}{1+e^{-(-0.11)}} \\ \dfrac{1}{1+e^{-0.36}} \end{Bmatrix} = \begin{Bmatrix} 0.4725 \\ 0.58904 \end{Bmatrix}$$

Step 5: The output of the output layer is found from $\{O\}_H$ and W matrix

$$\{I\}_0 = < 0.2 - 0.5 > \begin{Bmatrix} 0.4725 \\ 0.58904 \end{Bmatrix} = -0.20002$$

$$\{O\}_0 = \frac{1}{1+e^{-(-0.2002)}} = 0.45016 \quad \text{where } \lambda = 0.6$$

Step 6:

$$\text{Error} = (0.1 - 0.45016)^2 = -0.08667$$

$$l = (d - O_{01})\,(O_{01})\,(1 - O_{01}) \text{ where } d \text{ is the desired output} = 0.1$$

$$l = (0.1 - 0.12016)\,(0.45016)\,(1 - 0.45016) = -\,0.08667$$

$$[\Delta w]^1 = \lambda[\Delta w]^0 + \eta[y]$$

$$(0.6)\ \{0\} + <0.6> \begin{Bmatrix} -0.04095 \\ -0.05105 \end{Bmatrix} = \begin{Bmatrix} -0.02457 \\ -0.03063 \end{Bmatrix}$$

Step 7:

$$[y] = \{O\}_H l = \begin{Bmatrix} 0.4725 \\ 0.58904 \end{Bmatrix} < -0.08667 \geq \begin{Bmatrix} -0.04095 \\ -0.05105 \end{Bmatrix}$$

Step 8:

$$[g] = [W]\{l\} = \begin{Bmatrix} 0.2 \\ -0.5 \end{Bmatrix} (-0.08667) = \begin{Bmatrix} -0.017334 \\ -0.043335 \end{Bmatrix}$$

Step 9:

$$[l^*] = \begin{bmatrix} (-0.017334)\,(-0.04095)\,(1+0.04995) \\ (-0.043335)\,(-0.05105)\,(1+0.05105) \end{bmatrix} = \begin{bmatrix} 0.000738 \\ 0.002351 \end{bmatrix}$$

Step 10:

$$[\Delta V]^1 = \eta\{O\}_I\{l^*\} = (0.6)\left[\begin{Bmatrix} 0.5 \\ 0.8 \end{Bmatrix}\begin{bmatrix} 0.000738 \\ 0.001128 \end{bmatrix}\right] = \begin{bmatrix} 0.0002214 & 0.000705 \\ 0.0003542 & 0.001128 \end{bmatrix}$$

Step 11:

$$[W]^1 = [W]^0 + [\Delta W] = \begin{bmatrix} 0.2 \\ -0.5 \end{bmatrix} + \begin{Bmatrix} -0.02457 \\ -0.03063 \end{Bmatrix}$$

Step 12:

$$[V]^1 = [V]^0 + [\Delta V] = \begin{bmatrix} 0.1 & 0.4 \\ -0.2 & 0.2 \end{bmatrix} + \begin{Bmatrix} 0.0002214 & 0.000705 \\ 0.0003542 & 0.001128 \end{Bmatrix}$$

$$= \begin{bmatrix} 0.1002214 & 0.400705 \\ -0.199645 & 0.21128 \end{bmatrix}$$

4.4 SUMMARY OF BACK PROPAGATION ALGORITHM

1. Given a set of input -output pattern vectors.
2. Assume only one hidden layer and let the initial setting of weight be arbitrary.
3. Assume that input layer has linear units.
4. The output signal is the input activation value of each of these units.
   Assume a learning parameter.

5. Determine the activation of unit (i) in the input layer.
6. Determine the activation of unit (j) in the hidden layer.
7. The output signal from the jth unit on hidden layer can be calculated from the activation function.
8. Determine activation of unit (k) in the output layer and the corresponding outputs signal. Calculate the error term of this k^{th} output unit.
9. Update weights on output layers for k^{th}.
10. Error term for the jth hidden unit is to be formulated.
11. Update the weight on the hidden layer.
12. Calculate the error for the nth pattern.
13. Total error $E = \sum_{n-1}^{n} E_n$.
14. Apply the given patterns one-by-one several times in random order and Update the weights until the error is reduced to minimum.

4.5 KOLMOGOROV THEOREM AND OTHER FEATURES OF BACK PROPAGATION

According to this theorem, any continuous function of (n) variables can be computed by using only linear summations and non-linear continuously increasing function of only one variable.

$$F(X_1, X_2, X_3 \ldots\ldots\ldots\ldots X_n) \in R, \; x1[0, 1]$$

$$i = 1, 2, 3, \ldots\ldots n; \qquad \ldots(4.5.1)$$

This theorem implies that a three layers Perceptron with n ($2n + 1$) nodes using continuously increasing non-linearities can compute any continuous function of (n) variables. A 3-layer Perceptron network can be used to create continuous (likelihood) functions required in a classifier.

Activation function of type $v_p = \sum_{i=1}^{n} \lambda P \psi(x_c + x_{\in}) + P$ ...(4.5.2)

where λ is a real constant, $\in$ is a positive rational number, ψ is the continuously increasing function independent of F. The output layer has the activation function $y_j = \sum_{p=1}^{2n+1} g_i(V_P)$. The choice of g depends on F and $\in$ this theorem does not indicate how to find the nonlinear activation function y. However it confirms the feasibility of three layers network.

Another feature of interest is the number of training pairs. This depend on several factors from statistical observation, it is required to have at least 5 to 10 times as many training samples as the number of weights to be trained. If (q) is the number of training patterns and W is the number of weights to be trained. The necessary condition is $g > \frac{|w|}{1-a}$ where a = expected accuracy of test set. The sufficient condition is $g \geq \frac{|w|}{1-a} \log \frac{n}{1-a}$.

4.5.1 Reliability

1. For a large network it is possible that repeated training iterations improve the learning of the network but the resulting networks may perform poorly. This is called over training.
2. A solution is to monitor the error on the test set and terminate the training when the error increases (training may also be halted even when the network performance on the training set continues to improve).
3. To eliminate random fluctuations, the performance of the network test in monitored over several iterations, weight changes are computed on the basis of network performance on the training set with this stopping criterion, the final weights depends on the test data in an indirect manner.
4. A net with large number of nodes is capable of memorizing the training set at the expense of generalization. So, smaller nets are preferred over larger ones.
5. Another useful means is to improve by introducing noise in the training set.

4.5.2 Strictly Local Back Propagation

1. Back Propagation does not have the biological similarity regarding the operatives. Local Back Propagation overcomes these drawbacks. A strictly Local Back Propagation net has 3 cortical units, synaptic units and thalamic units.
2. A Cortical unit sums inputs and sends the results as signal to the next unit.
3. The function of a synaptic unit is to receive a single input from the cortical unit, apply an activation function, multiply the result a weight and pass the result to single thalamic unit.

4.6 LEARNING DIFFICULTIES AND IMPROVEMENTS

4.6.1 Factors of Importance

(*a*) **Choice of Learning Coefficient:** The learning coefficient η has to be between 0 and 1. Firstly it cannot be negative as change of weights would shift from ideal position. If it is zero, no learning takes place. If it is 2, the network becomes unstable. For any value of $n > 1$, there will be oscillation. Larger valve are needed when input patterns are close to ideal. The governing equation is

$$\eta = \frac{1.5}{(N_1^2 + N_2^2 + ... N_p^2)} \qquad ...(4.6.1)$$

Where N_1 is the number of pattern of type 1 and N_p is the number of patterns of type p. For $n > 0.5$, the weights are changed fast whereas for $n < 0.2$ the weights are changed in small increment. The optimum value is 0.6 in many cases.

(*b*) **Momentum coefficient *L*:** The momentum term is seen from the equation

$$[\Delta W]^{nth} = -\eta \frac{\partial E}{\partial W} + L[\Delta W]^n \qquad ...(4.6.2)$$

Basically L enhances the training process. Momentum coefficient is used to overcome the effect of local minima and reach the global minimum. A "second order" Back Propagation uses a second derivative for more accurate estimation (Parker '87).

(*c*) **Number of Hidden Nodes:** The criterion is to select the nodes which would not disturb the performance by keeping the memory requirements to minimum. The number of separable regions M is supposed to be a $H - 1$.Where H is the number of hidden nodes $H = K - 1$

according to Huang and Huang where K is the number of elements in the learning set. For generalization, the ratio of number of hidden nodes to patterns should be 1:10. According to V_c dim of probability theory, if training samples are T (> V_c dim), l_1, l_2 and l_3 correspond to input, hidden and output nodes then

$$l_2 = \frac{10T}{l_1 + l_3}$$

(*d*) **Sigmoidal gain:** In the equation $\lambda = \frac{1}{1+e^{-\lambda x}}$ where λ indicates that as the scaling factor is increases, the input output characteristics approaches a point where activation function reaches the saturation. As the scaling factor is increased learning rate and momentum factor have to be decrease to avoid oscillations.

(*e*) **Threshold valve:** In the equation $\theta = \frac{1}{1+e^{-\lambda(I+\theta)}}$, where θ is the Threshold or bias or noise factor . In one method θ is kept small and not changed during training. In another method it is chosen at some random value which is not changed.

(*f*) **Simulated Annealing:** For solving the problem of local minima, a parameter called temperature is introduced so that for higher temperature more randomness can be introduced. Generic algorithm is used in this case by starting with multiple initial weights and recombining trained weights during training.

4.6.2 Limitation/Difficulties of Back Propagation

The most important limitation is the slow convergence. Although performance of this technique in practice is acceptable, there is no proof of convergence.

Because of stochastic gradient descent on a non-linear error surface, the result is likely to converge to some local minimum. There is no easy techniques to overcome this defect (stochastic learning algorithms are available ofcourse).

The problem of scaling is another drawback. When the complexity of the problem increases, convergences as well as generalisation are affected. Effects of scaling can be handled with prior information. Modular architecture also helps in avoiding this problem.

When the desired output is not, known Back Propagation learning cannot be applied. Reinforcement learning based on information on correctness of response helps to overcome this problem.

4.6.3 Special Improvements

In an improved technique of Back Propagation, convergence towards global minimum is achieved if the information from the given training patterns are used more effectively. In a technique of this type, supervised learning problem is posed as a non-linear system identification problem [Haykin '91]. This results in a learning algorithm called Kalman type learning based on piece wise linear approximation of optimal filtering.

Improvement is also possible if the supervised learning is used as an unconstrained optimisation problem, where error function is taken as the cost function [Battilti '92].

The posteriori class probabilities of multilayer feed forward network can be used as the desired response in a learning algorithm based on information measure for the cost function [Haykin '94].

Fuzzy representation of input and output can also be used for improvement of Back Propagation.

4.6.4 Applications

Main applications are:

(*a*) Realisation of Logic functions.
(*b*) Pattern Classification.
(*c*) Pattern Mapping.
(*d*) Function Approximation.
(*e*) Estimation of probability distribution and prediction.

Using these, real world tasks such as speech/character recognition, system identification, passive sonar detection the classification, and speech synthesis can be performed.

Examples

(*a*) Design of journal bearing.
(*b*) Classification of soil.
(*c*) Hot extrusion of steel.

SUMMARY

- This unit focuses on the interconnections of neurons in the network modes.
- The concept of learning rules and postulates has been presented along with suitable examples.
- A portion of neural dynamics has been touched upon.
- The other learning techniques have also been discussed.
- Generalized delta rule and back propagation concept have been introduced

Short Questions

1. What are the types of neural networks?
2. What is the main difference between supervised and unsupervised learning?
3. Explain Hebbian, Neo-Hebbian and Kohonen learning?
4. Explain how fuzzy learning brings closeness to biological systems?
5. Briefly state the perceptron learning rule?
6. What is co-relational learning?
7. What is the meaning of back propagation?
8. What is LMS rule?
9. Explain difference between instar and outstar?
10. How is pattern recognition done?

❑❑❑

5

ASSOCIATIVE MEMORY

5.0 INTRODUCTION

The human mind remembers and recognizes by method of association. When we see a bridge, we know that it is so because in our mind we have a picture of bridge stored already and we associate the object seen with the object stored and conclude. Associative memories are based on such principles. Variations of auto association, hetero association, bidirectional association, and Hopfield type of association are the topics of this unit. The main utilization of these is in pattern recognition. Actually humans see everything as a pattern where as machines see things as data. Humans can understand but machines can only recognize. Patterns can be stored, new pattern can be presented and old patterns can be recalled. Classification of patterns is to retrieve the class label once the new pattern is presented. In Pattern Mapping given a set of input and output patterns, objective is to capture their relationship so that when a test pattern is given a suitable pattern can be retrieved to identify the subset of patterns grouping or clustering done in feature mapping.

In the context of such varied application this unit also covers various types of memories such as content addressable and instance based.

5.1 PARADIGM OF ASSOCIATIVE MEMORY

The block diagram shown is that of an associative memory which associates an input vector X into an output vector V by mapping. We may say that the system basically maps vectors x to vectors V. (Input in the pattern space R^n and output in the pattern space R^m.) The matrix transformation is $V = M[X]$, where M is the general non-linear matrix (operation) and changes its meaning for each type of memory model. The structure of M defines the neural memory paradigm. In case of Dynamic memories. M has a time variable such that the output is available at sometime after the presentation of the input.

Fig. 5.1

5.1.1 Vectorial Rules

1. For a given model of memory, M is expressed in terms of given prototype vectors which must be stored.
2. The algorithm permitting computation of M is called storage algorithm.
3. The mapping done in the equation $V = M[X]$ performed on a key vector X is called Retrieval (Retrieval may or may not be effective for providing a desired solution.)
4. Storage algorithm can be formulated using one or two sets of prototype vectors this depends on the type of association.

The structure of M represents a specific neural memory paradigm. For dynamic memories $\overline{V}$ is available at the output at some time after the application of input. The mapping as in equation $\overline{V} = M(\overline{U})$ represents the process of retrieval which may or may not provide a desired solution prototype/undesired prototype. It may not even provide a stored prototype at all. In these extreme cases erroneously recalled output does not pertain to the prototype.

Let us assume that there are P stored pairs of association defined as

$$u^{(i)} \rightarrow v^{(i)} \text{ for } i - 1,2,3 \ldots \ldots .. p$$

$$v^{(i)} \rightarrow u^{(i)} \text{ for } i - 1,2,3 \ldots \ldots .. p$$

Then the network can be called hetero associative memory. The association of pairs of ordered sets of vector $u^{(i)}$ and $v^{(i)}$ is hetero associative memory.

In case of auto association, $u^{(i)} \rightarrow v^{(i)}$, $v^{(i)} \rightarrow u^{(i)}$ for $i = 2 \ldots \ldots p$

5.1.2 Associative Memory Vs Digital Memory

In digital computer memory data are accessed when the corresponding addresses are given. The binary address is activated by the address decoder. This configuration (Fig. 5.2) belongs to address – addressable memory U_i.

Associative memories on the other hand are content addressable. Words are accessed based on content of key vector $\overline{u}$ when a portion of $u^{(i)}$ $i = 1, 2, \ldots p$ excites the network then a complete version of $\overline{u}$ is generated. In hetero-associative memory, content of $u^{(i)}$ provides the stored response $u^{(i)}$ (it may be noted that there is no storage for prototype $u^{(i)}$ or $v^{(i)}$ at any location within the network.

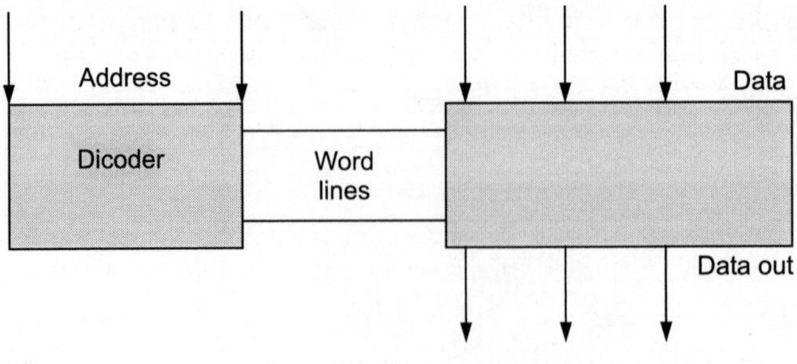

Fig. 5.2

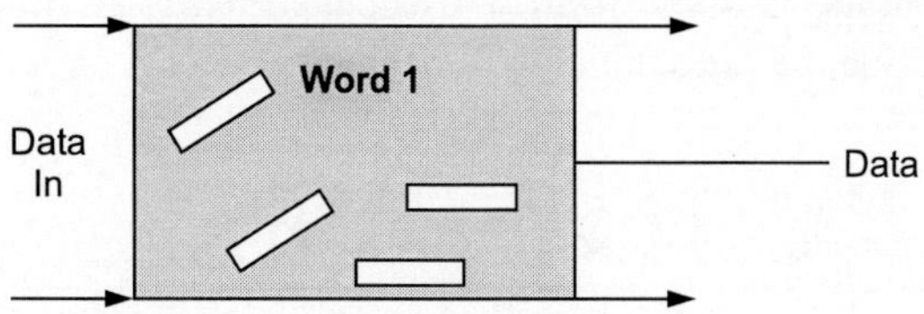

Fig. 5.3 *Shows the Underlying Concepts.*

A special category of auto association is that of Hopfield auto associative memory [Fig. 5.4] in this case $u^0 = v^0$ for initialization and input is removed thereafter.

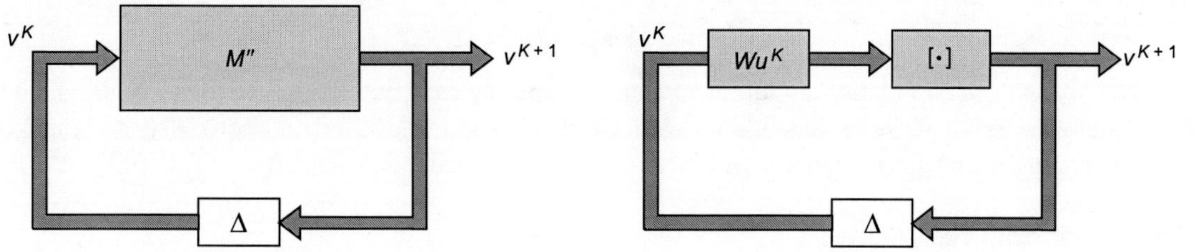

(*a*) *Basic block diagram* (*b*) *Expanded block diagrams*

Fig. 5.4

The vector retrieved at instant *K* is computed with this initial state

$$v^{(k+1)} = M''[u^k] \qquad ....(5.1.0)$$

Figure 5.4(*a*) shows Hopfield auto associative memory without initializing the input. Operator M^{11} consists of multiplication of a weight matrix followed by ensemble of non-linear mapping operations of layer of neurons.

The last equation can rewritten as

$$v^{(k+1)} = [W u^k] \qquad ...(5.1.1)$$

The operator $\overline{|[.]}$ is a non-linear matrix operator with special diagonal elements

$$\begin{aligned} \text{sgn}(\cdot) &= 0 \ \dots \ \dots \ \dots \ \dots \ 0 \\ [\cdot]\ 0 &= \text{sgn}(\cdot) \ \dots \ 0 \\ 0 &= 0 \ \dots \ \text{sgn}(\cdot) \end{aligned} \qquad ...(5.1.2)$$

5.1.3 Pictorial View of Auto Association and Hetero Association

Let a set of patterns be stored in a network (Fig. 5.5) when the network is presented with a pattern similar to a member of the stored set, it may associate the input with closest stored pattern. For example, when we see a tower or a part of it we can recognize it as a tower because we can associate it with the pattern stored in our memory by way of experience or learning. This process is termed AUTO ASSOCIATION. The distorted input pattern can serve as a reminder for forming the original pattern.

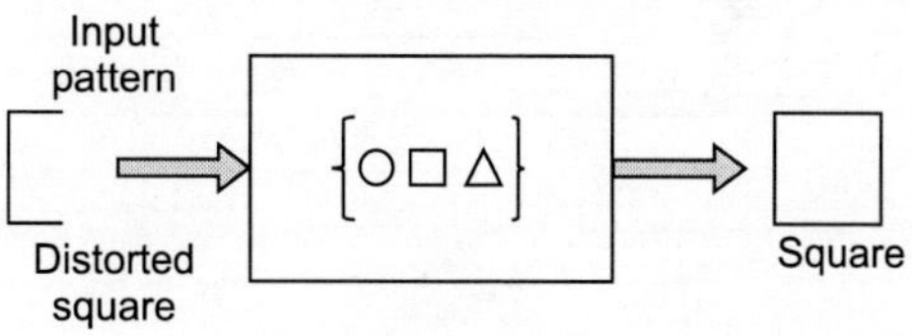

Fig. 5.5

When the patterns stored are in some form of associated pairs, an input result in retrieval of the pattern associated with it inside the network (Fig. 5.6) this process is called hetero association.

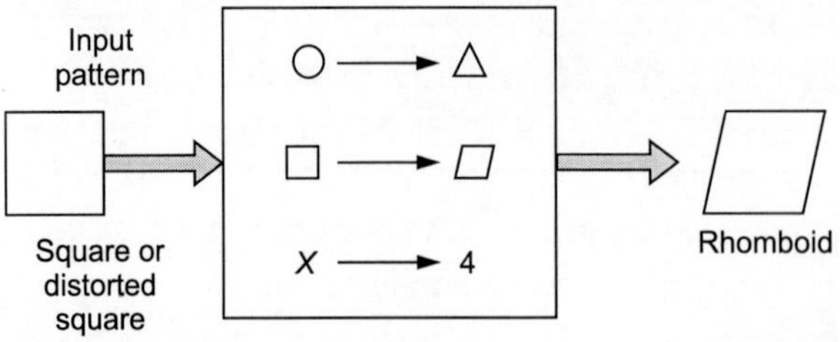

Fig. 5.6

The process of computation of output for a given input by such networks is termed Recall which is the proper processing phase for a neural network (its basic is to retrieve information). This corresponds to the decoding of the stored content which may have been encoded previously.

5.1.4 Associative Memory

Associative Memory is a storehouse of associated patterns which are encoded in some form when the storehouse is excited by a pattern the associated pattern is recalled as the output.

Auto associative memories are used for image refinement while hetero associative memories are used for association of patterns. Auto associative correlation memories are called auto correlates and hetero associative correlation memories are known as hetero correlates.

5.2 PATTERN MATHEMATICS

Based on the principle of recall associative memory can be static or dynamic. Basically an associative memory can be thought of as a mapping between pattern spaces R^m and R^n so that for a general matrix type mapping patterns, $X\ \overline{u} \in R^m,\ \overline{v} \in R^n$; we can write

$$\overline{v} = M(\overline{u})$$

The operator M can have different forms for different models of memory. The algorithm for M is the recording or storage algorithm and it is based usually on input pattern vectors.

Static networks recall an output when an input is given in one feed forward pass where as dynamic networks recall through time taking input – output feedback mechanism.

The static networks are non-recurrent and dynamic networks are recurrent (Fig. 5.7).

The recursive equations are used until equilibrium is reached in the dynamic network.

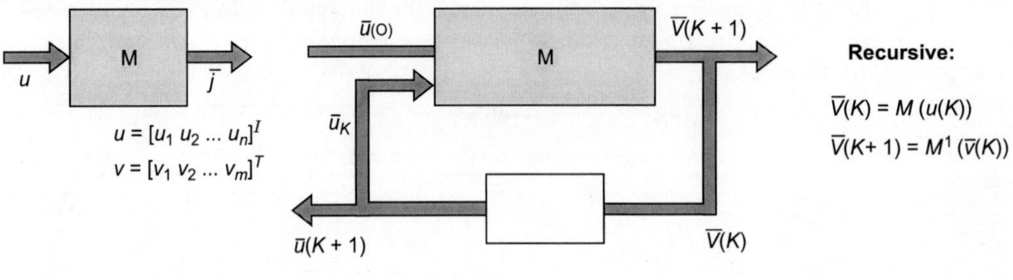

Fig. 5.7

5.3 HEBBIAN LEARNING AND ASSOCIATIVE MATRIX

In first order auto correlates based on Hebbian learning the association (also called connection matrix) is given by

$$T = \sum_{i=1}^{m} [A_i^T][A_i]$$

where A_1, A_2 .... A_m are the bipolar patterns.

The recall equation is given by

$$a_j^{new} = f(a_i, t_{ij}, a^{old}) \ \forall_j = 1,2 \ \dots \dots \ p, \qquad \dots(5.3.0)$$

where

$$A_i = (a_1, a_2 \ \dots \dots \dots \ a_p) \text{ and}$$

$$f(\alpha, \beta) = \begin{cases} 1 & if \quad a > 0 \\ \beta & if \quad \alpha = 0 \\ -1 & if \quad a < 0 \end{cases} \qquad \dots(5.3.1)$$

It may be noted that this recall equation is analogous to working of human recall.

5.3.1 Hebbs Rule for Patterns

For associative memories using neural network, Hebb's rules gives simple methods of design the basic algorithm is

Step 1 : Initialize all weights

Step 2 : For each training input, target output, execute Steps 3-5

Step 3 : Activation for input are to be determined ($i = 1, \dots n$)

Step 4 : Activation for output units to current target output ($i = 1 \dots m$)

Step 5 : The weight to be adjusted

$$w_{ij}(new) = w_{ij}\,(old) + x_i y_i$$

If training vector is S, testing vector is T, target vector is t

$$S = (S_1 \dots S_i, S_n) \text{ and } T = (t_1 - t_j, t_m)$$

Matrix product ST gives the weight matrix to be stored

For a set of association $S(P) : t(p)P = 1, \dots \dots p$

$$w_{ij} = \sum_{p=1}^{p} S_i(p) t_j(p). \quad \text{(sum of outer product)} \qquad \dots(5.3.3)$$

The weight determination can be written as

$$w_{ij} = \sum_{p=1}^{p} S^t{}_p(t)(p). \qquad \dots(5.3.4)$$

5.3.2 Delta Rule of Pattern Association

The adjustment to a synaptic weight is proportional to the product of the error signal and the input signal of the synapse. This rule can be applied for linearly dependent inputs and it assumes that error signal is directly measurable.

For Single output unit $\Delta w_{ij} = \alpha(t - y)$ ... ...(5.3.5)

For several output units $\Delta w_{ij} = \alpha(t_1 - y_1)x_i$ ... ...(5.3.6)

Here i^{th} input unit and j^{th} output unit are considered.

Extended Delta Rule

In case of arbitrary, differentiable activation function, the weight update y is

$$\Delta w_{ij} = \alpha(t_1 - y_1)x_i f'(y_{inj})... \quad ...(5.3.7)$$

Squared error for particular training pattern is

5.3.3 Hetero Associative

In this case the net finds an appropriate output vector which corresponds to an input vector which can be either one of the stored patterns or patterns which may or may not be identical

Step 1 : The weights are initialized used Hebb (or Delta) rule.

Step 2 : For each input vector execute Steps 3 to 5.

Step 3 : Activation of input layer units are determined.

Step 4 : The net input to the output is computed.

Step 5 : The activation of the output above is determined.

$$F(y) = \begin{matrix} 1 \\ 0 \\ -1 \end{matrix} \quad \text{depend} \quad \begin{matrix} y > 0 \\ y\,;\, y = 0 \\ y < 0 \end{matrix} \quad ...(5.3.8)$$

The output vector y gives the pattern associated with the input vector x. The other activation function can be adapted and if the target response of the net is binary, suitable activation function,

$$f(x) = \begin{matrix} 1 \\ 0 \end{matrix} \text{depends on} \begin{matrix} x > 0 \\ \text{or } x \le 0 \end{matrix} \text{...... may be used} \quad ...(5.3.9)$$

5.3.4 Modification of Data

1. Binary input and target vectors are converted to bipolar representation for the construction of weight matrix. The input vector used in testing and response of the net may be in binary form to give a hybrid data representation.
2. All the vectors (training input/ target output, testing input and the response are expressed bipolar form.

EXAMPLE 5.1: A Hetero associative network is defined as per the following:

$S_1 = 1100 \quad S_2 = 0100 \quad S_3 = 0011 \quad S_4 = 0010$

$t_1 = 10, \quad t_2 = 10, \quad t_3 = 01, \quad t_4 = 01$

Solution. We have to find the weight matrix and test the network.

$$w_1 = S_1^t t_1 = \begin{bmatrix} 1 \\ 1 \\ 0 \\ 0 \end{bmatrix} [1 \;\; 0] = \begin{bmatrix} 10 \\ 10 \\ 00 \\ 00 \end{bmatrix}$$

$$w_2 = S_2^T t_2 = \begin{bmatrix} 0 \\ 1 \\ 0 \\ 0 \end{bmatrix} = \begin{bmatrix} 00 \\ 10 \\ 00 \\ 00 \end{bmatrix}, \; w_3 = S_3^T t_3 = \begin{bmatrix} 00 \\ 00 \\ 01 \\ 01 \end{bmatrix}, \; w_4 = S_4^T t_4 = \begin{bmatrix} 00 \\ 00 \\ 01 \\ 00 \end{bmatrix}$$

$$w = w_1 + w_2 + w_3 + w_4 = \begin{bmatrix} 10 \\ 20 \\ 02 \\ 01 \end{bmatrix}$$

For the first training vector input pattern is $X_1 = [1\ 1\ 0\ 0]$

$$y_j = \sum_{i=1}^{4} x_i w_{ij} \,/\, y_{ji} = x_1 w_{11} + x_2 w_2 + x_3 w_3 + x_4 w_{41}$$

$$= 1 \times 1 + 1 \times 2 + 0 \times 0 + 0 \times 0 = 3$$

$$Y_j^2 = 1\ f(y_j^2) = 0$$

the response is correct. Similarly it can be shown that response of all the input pattern are correct.

For the cases, when the input vector is not similar to the training input, we can follow similar approach and starting from the calculation weight matrix distance Y_j^1 and y_j^2 and then the response.

EXAMPLE 5.2: For the hybrid (binary/ bipolar) form of Hebb's rule we have to find the weighted matrix for the network given by:

$S(1) = 100$	$S(2) = 0100$	$S(3) = 0011$	$S(4) = 0010$
$T(1) = 10$	$t(2) = 10$	$t(3) = 01$	$t(4) = 01$

Solution: W_{ij} is given by

$$W_{ij} = S(2s_i(p) - 1)\ (2t_j(p) - 1)$$

$$W_{11} = (2s_1(1)-1)\ (2t_1(1)-1) + (2S_1(2)-1)\ (2t1(2) - 1)$$
$$+ (2S_1(3) - 1)\ (2t_1(3) - 1) + 2(S_1(4) - 1)\ 2t_1(4) - 1)$$
$$= (1)\ (1) + (-1)\ (1) + (-1)\ (-1) + (-1)\ (-1) = 2$$

$$W_{12} = (2S_1(1) - 1)\ (2t_2(1) - 1) + (2S_1\ (2) - 1)\ (2t_2(2) - 1) +$$
$$(2S_1(3) - 1)\ (2t_2(3) - 1) + (2S_1(4) - 1)(2t_2\ (3) - 1) = -2$$

$W_{21} = 4,\quad W_{22} = -4$

$W_{31} = -4\quad W_{32} = 4$

$W_{41} = -2\quad W_{42} = 2.$

5.3.5 Hamming Distance

To understand the extent meaning of hamming distance it is pertinent to study the well known numerical example for working of auto correlate.

Let us consider that pattern A_1 (–1,1 –1,1)

A_2 = (1, 1, 1, –1), A_3(–1, –1, –1, 1) are stored.

The connection matrix $= \sum_{i=1}^{3} [A_1]^T [A_1]$

$$= [A_1]^T [A_1] + [A_2]^T [A_2] + [A_3]^T [A_3]$$

$$= \begin{bmatrix} 3 & 1 & 3 & -3 \\ 1 & 3 & 1 & -1 \\ 3 & 1 & 3 & -3 \\ -3 & -1 & -3 & 3 \end{bmatrix}$$

If the auto correlator is given an input of A_2

$$d_1^{new} = f(3 \times 1 + 3 \times 1 + 3 \times 1 + (-1)(-1), 1) = 1$$

$$d_2^{new} = f(6, 1) = 1 \quad d_3^{new} = f(0,1) = 1, \quad d_4^{new} = -1$$

So the retrieved pattern is same as the input pattern. Now, if a distorted pattern is put such as we have to find the proximity of the noisy vector to the stored one using a measure called Hamming Distance which can be expressed as

$$HD\,(a,b) = \sum_{i=1}^{n} |a_i - b_i| \quad (a \text{ and } b \text{ are vectors having } n \text{ components})$$

For the benefit of the student the detailed matrix calculation are shown in the following.

EXAMPLE 5.3: The patterns A_1 = (-1, 1, –1, 1), A_2 = (1, 1, 1 -1) A_3 = (–1, –1, –1, 1) are stored in the auto correlate.

Solution:

$$T = \Sigma[A_i^T][A_t] = \begin{bmatrix} (-1)\times(-1) & 1\times(-1) & (-1)\times(-1) & 1\times(-1) \\ (-1)\times 1 & 1\times 1 & (-1)\times 1 & 1\times 1 \\ (-1)\times(-1) & 1\times(-1) & (-1)\times(-1) & 1\times(-1) \\ (-1)\times 1 & 1\times 1 & (-1)\times 1 & 1\times 1 \end{bmatrix}$$

$$+ \begin{bmatrix} 1\times 1 & 1\times(-1) & 1\times 1 & (-1)\times 1 \\ 1\times 1 & 1\times 1 & 1\times 1 & (-1)\times 1 \\ 1\times 1 & 1\times(-1) & 1\times 1 & (-1)\times 1 \\ 1\times(-1) & 1\times 1 & 1\times(-1) & (-1)\times(-1) \end{bmatrix} + \begin{bmatrix} (-1)\times(-1) & (-1)\times(-1) & (-1)\times(-1) & (-1)\times(-1) \\ (-1)\times(-1) & (-1)\times(-1) & (-1)\times(-1) & (-1)\times(-1) \\ (-1)\times(-1) & (-1)\times(-1) & (-1)\times(-1) & (-1)\times(-1) \\ (-1)\times(-1) & (-1)\times 1 & (-1)\times 1 & (-1)\times 1 \end{bmatrix}$$

$$= \begin{bmatrix} 1 & -1 & 1 & -1 \\ -1 & 1 & -1 & 1 \\ 1 & -1 & 1 & -1 \\ -1 & 1 & -1 & 1 \end{bmatrix} + \begin{bmatrix} 1 & 1 & 1 & -1 \\ 1 & 1 & 1 & -1 \\ 1 & 1 & 1 & -1 \\ -1 & -1 & -1 & -1 \end{bmatrix} + \begin{bmatrix} 1 & 1 & 1 & -1 \\ 1 & 1 & 1 & -1 \\ 1 & 1 & 1 & -1 \\ 1 & -1 & -1 & -1 \end{bmatrix} = \begin{bmatrix} 3 & 1 & 3 & -3 \\ 1 & 3 & 1 & -1 \\ 3 & 1 & 3 & -3 \\ -1 & -1 & -3 & 3 \end{bmatrix}$$

When a stored pattern A_2 = (1, 1, + 1, –1) is presented $a^{old} = 1$

$$a_1^{new} = f(1\times 3 + 1\times 1 + 1\times 3 + \overset{(-1)\times(-3)}{3}, 1) = 1$$

$$a_2^{new} = f(\overset{6}{1\times 1 + 3\times 1 + 1\times 1} + -1\times 1, 1) = 1$$

$$a_3^{new} = f(1\times3+1\times1+3\times1+3\times1, 1) = 1$$

$$a_4^{new} = f(-1\times1+1\times1 \overset{-10}{+} -3\times+1+3\times-1, 1) = -1$$

So the new pattern = (1 1 1 –1) = A_2

Similar for A_3 it can be shown that A_3 = (–1, –1, –1, 1)

$a_1^{new} = -1,\quad a_2^{new} = -1,\quad a_3^{new} = -1 \quad a_4^{new}$ = By expression

$$a_1^{new} = f(-3+-1 \overset{-10}{+} -3\pm3, 1=-1)$$

$$\left.\begin{array}{l} a_2^{new} = f(-1-3-1-1, 1) = -1 \\ a_3^{new} = f(-3-1-3-3, 1) \\ a_4^{new} = f(3+1+3+3, 1) = 1 \end{array}\right\} A_3$$

5.3.6 Computation of Hamming Distance

Let a distorted vector A_1 = (1, 1, 1, 1) be presented to calculate the proximity of the vector, Hamming distance is defined as

$$HD(a, b) = \sum_{i=1}^{n} |a_i - b_i| \ldots\ldots\ldots \qquad \ldots(5.3.9)$$

$$HD(A^1, A_1) = (2+0+2+0) = 4$$

$$HD(A^1, A_2) = 0+0+0+2 = 2$$

$$HD(A^1, A_3) = 2+2+2+0 = 6$$

Thus A_1 is closest to A_2 and so it can be treated as a noisy version of A_2

$$\left.\begin{array}{l} a_1^{new} = f(4, 1) = 1 \\ a_2^{new} = f(4, 1) = 1 \\ a_3^{new} = f(4, 1) = 1 \\ a_4^{new} = f(-4, 1) = 1 \end{array}\right\} A_2$$

Thus even in case of partial vectors the auto correlator results in restoration and refinement of the pattern or removal of the noise in order to retrieve the closest matching pattern stored.

5.3.7 Hetero Correlators: Kosko's Discrete BAM

1. Kosko extended unidirectional Auto association to bidirectional process.
2. Discrete BAM has the ability of recalling stored pairs part in presence of noise.
3. There are N training Pairs $\{(A_1, B_1), (A_2, B_2), .. (A_i, B_i), (A_n, B_n)\}$

Where $A_i = (a_{i1}, a_{i2} \ldots\ldots\ldots (a_{in})$

$B_i = (b_{i1}, b_{i2} \ldots\ldots (b_{ip})$

A_{ij} and b_{ij} are either in ON or OFF stage

4. ON = 1/OFF = 0 or bipolar ON = 1 and OFF = –1

Correlation Matrix $$M = \sum_{i=1}^{N} X_i^t Y_i \qquad \ldots(5.3.10)$$

Given a training pair we have to retrieve the output start with bipolar and $A_i = 1$ and apply a sequence of condition to get (α', β'); (α'', β'') such that one of the β values a finite sequence $\alpha^1, \beta^1) (\alpha^{11}, \beta^{11})$ until an equation pt (α_F, β_F) matches with vector β_i.

Steps:

1. $X_i = A_i = \alpha$
2. β' = hard limiting function of M.
3. α' = hard limiting function of $\beta' MT$.
4. β'' = hard limiting function of α'M
5. α'' = hard limiting function of β''
6. β'' = hard limiting function of α''

EXAMPLE 5.4: For $N = 3$

A_1 = (100001) B_1 = (11000)

A_2 = (011000) B_2 = (10100)

A_3 = (001011) B_3 = (01110)

X_1 = (1–1–1–1–11) Y_1 = (11–1–1–1)

X_2 = (–111–1–1–1) Y_2 = (1–11–1–1)

X_3 = (–1–11–111) Y_3 = (–111–1)

Solution: We have converted this to bipolar logic and got the connection matrix.

$$M = X_1^T Y_1 + X_2^T Y_2 + X_3^T Y_3$$

$$\begin{bmatrix} 1 \\ -1 \\ -1 \\ -1 \\ 1 \end{bmatrix} \underset{X_1^T Y_1}{[1 \;\; 1 \;\; -1 \;\; -1 \;\; -1]} = \begin{bmatrix} 1 & 1 & -1 & -1 & -1 \\ -1 & -1 & 1 & 1 & 1 \\ -1 & -1 & 1 & 1 & 1 \\ -1 & -1 & 1 & 1 & 1 \\ -1 & -1 & 1 & 1 & 1 \\ 1 & 1 & -1 & -1 & -1 \end{bmatrix}$$

$$\begin{bmatrix}-1\\1\\1\\-1\\-1\end{bmatrix}[1\ \ -1\ \ 1\ \ -1\ \ -1] = \begin{bmatrix}1 & -1 & -1 & -1 & 1\\1 & -1 & -1 & -1 & 1\\-1 & +1 & -1 & -1 & 1\\+1 & -1 & -1 & -1 & 1\\-1 & 1 & 1 & 1 & -1\\-1 & 1 & 1 & 1 & -1\end{bmatrix}$$

$$M = \begin{bmatrix}1 & -1 & -3 & -1 & 1\\1 & -3 & 1 & -1 & 1\\-1 & -1 & 3 & 1 & -1\\-1 & -1 & -1 & 1 & 3\\-3 & 1 & 1 & 3 & 1\\1 & 3 & -1 & 1 & -1\end{bmatrix}$$

EXAMPLE 5.5: We have to retrieve the pattern corresponding to X_3

Solution: We Start with $\propto = X_3$ hoping to retrieve Y_3

I. $\alpha M = (-1\ -11\ -\ 111)\ (M)$

$(-1 \times 1) + (-1 \times 1) + (-1 \times 1) + (-1 \times -1) + (-3 \times 1) + (1 \times -1)$

$= -1\ -1\ -1 + 1 - 3 - 1 = -6$

II. $\alpha XM = (-\ 6666\ -\ 6)$

III. $b' = \text{HLF}\ (\ \alpha M) = (-1\ 1\ 1\ 1{-1})$ Hard Limiter function

$\beta^1 M^T = (-\ 5 - 55 - 375\ )$

$$HLF\ (\beta'\ M^T\ ) = (-1\ -11\ -111) = \alpha^1$$

IV. $\alpha' M = (-\ 6666\ -\ 6)$

$$\varphi\ (\alpha' M)\ = \beta^2 = (-1111\ -\ 1) = \beta'$$

$$\beta'\ = Y_3.\ \text{Hence}\ (\alpha_F, \beta_F) = (X_3, Y_3)$$

5.3.8 Linear Associator

1. Conventional associative memories are feed forward instantaneous. The basic task is to learn the association with in p vector pairs $\{\ U(i), V(i)\}$ for $i = 1,2\ \ldots\ldots\ p$ for the linear associative memory input pattern $\overline{U}$ is presented and mapped to output by.

$$V\ = W\ U$$

2. Figure 5.8 shows a linear associative network. Using neurons, mapping can be written as

$$V\ = M_1[Wu]$$

where $M_1[\cdot]$ is a dummy linear matrix operator of mXm unity matrix.

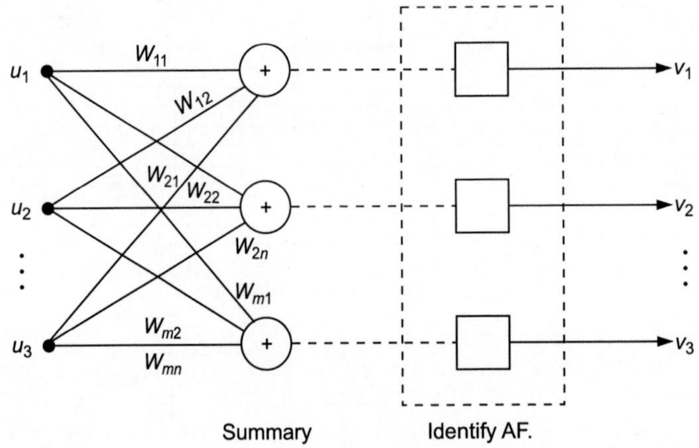

Fig. 5.8

Let p associations be stored in the linear association the pairs of vectors $\{X^{(i)}, f^{(i)}\}$ for $i = 1,2 \ldots.p$ called stimuli – forced responses are to be taken

$$X^{(i)} = [X_1^{(t)}\ X_2^{(t)} \ldots\ldots X_m^{(i)}]^T \qquad \ldots(5.3.11)$$

$$F^{(i)} = [F_1^{(t)}\ F_2^{(t)} \ldots\ldots F_m^{(i)}]^T \qquad \ldots(5.3.12)$$

The objective of linear associator is to implement the mapping

$$f^{(i)} + n_2 = Wx^{(i)}$$

or

$$x^{(i)} \rightarrow f^{(i)} + n^i \text{ for } i = 1,2\ldots\ldots.\ P \qquad \ldots(5.3.13)$$

Here n_i is the noise factor which has to be minimized using Hebbian Learning rule, updating for ith output node and j^{th} input node is

$$W_{ij}^i = W_{ij} + f_i x_j$$

$$\Rightarrow \qquad W' = W + f \times T \qquad \ldots(5.3.14)$$

Initializing the weights in their unbiased position $W_0 = 0$, the outer product learning rule is

$$W_{ij} = f^{(\text{i})} s^{(\text{i})} T$$

As there are p pairs to be trained $W' = \Sigma_{i=1}^{p} f^{(i)} X^{(i)T}$ ...(5.3.15)

The matrix W' has the form of Cross-Correlation Matrix when we have to perform an associative recall of the vector $f^{(i)}$ (as $x^{(\text{i})}$ is applied as stimulus) we can $x^{(j)}$ as a key vector at the input the retrieval formula is

$$v = (\Sigma f^{(i)}\, x^{(i)\,T})\, x^0) \qquad \ldots(5.3.16)$$

$$= f^{(i)}\, x^{(i)T} x^{(j)} + - - - - + f^{(j)} x^{(j)T} x^{(j)} + - - - f^{(p)} x^{(p)} x^{(j)}$$

For ideal mapping with no noise term,

$$V = f^{(j)} \text{ hence}$$

$$x^{(i)T}\, x^{(j)} = 0 \text{ for } i \neq j$$

$$x^{(i)T}\, x^{(j)} = 1$$

Interesting property of the linear associator is the case of auto associative operation with p distinct n-dimensional prototype patterns $x^{(I)}$ this network is called an auto correlator by having $f^{(i)} = x^{(i)}$ the autocorrelation matrix W^T is given by following

$$W^T = \sum_{i=1}^{p} f^{(i)} x^{(i)T} \quad \text{...(5.3.17)}$$

For distorted pattern of (j) is the distortion then error take rise is

$$\sum_{i \neq j}^{p} (f^{(i)} x^{(i)})^T \Delta^{(j)} \quad \text{...(5.3.18)}$$

5.3.9 Bidirectional Associative Memory Vs Linear Association

Kosko (1988) extended the Hopfield model by incorporating an additional layer to perform recurrent auto associations as well as hetero associations on the stored memories.

The network structure of the Bidirectional Associative Memory model is similar to that of the linear associator but the connections are bidirectional, i.e., $w_{ij} = w_{ji}$, for $i = 1, 2, ..., m$ and $j = 1, 2,..., n$. Also, the units in both layers serve as both input and output units depending on the direction of propagation. Propagating signals from the X layer to the Y layer makes the units in the X layer act as input units while the units in the Y layer act as output units. The same is true for the other direction, i.e., propagating from the Y layer to the X layer makes the units in the Y layer act as input units while the units in the X layer act as output units. Given below is an illustration of the BAM architecture.

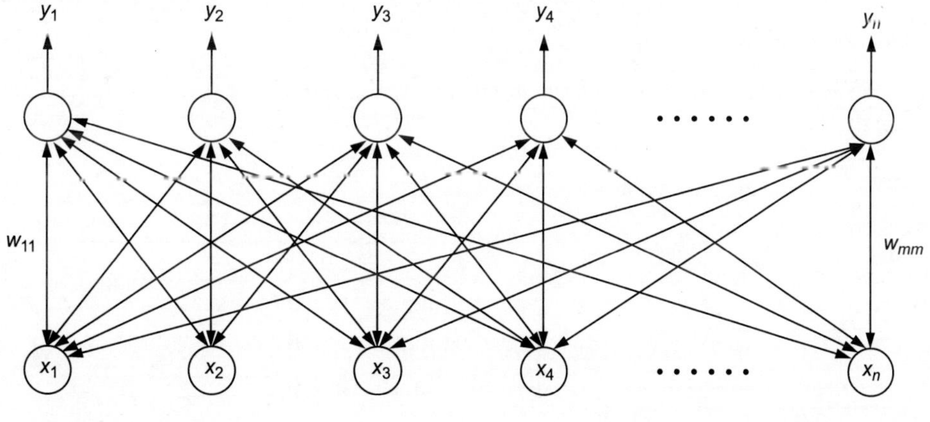

Fig. 5.9

5.3.10 BAM Model

Just like the linear associator and Hopfield model, encoding in BAM can be carried out by using:

$$W^k = X_k^T Y_k$$

to store a single associated pattern pair and

$$W = \propto \sum_{k-1}^{p} w_k$$

to simultaneously store several associated pattern pairs.

After encoding, the network can be used for decoding. In BAM, decoding involves reverberating distributed information between the two layers until the network becomes stable. In decoding, an input pattern can be applied either on the *X* layer or on the *Y* layer. When given an input pattern, the network will propagate the input pattern to the other layer allowing the units in the other layer to compute their output values. The pattern that was produced by the other layer is then propagated back to the original layer. The units in the layer are allowed to compute their output values. The new pattern that was produced by the original layer is again propagated to the other layer. This process is repeated until further propagations and computations do not result in a change in the states of the units in both layers where the final pattern pair is one of the stored associated pattern pairs. The final pattern pair that will be produced by the network depends on the initial pattern pair and the connection weight matrix.

Several modes can also be used to update the states of the units in both layers namely synchronous, asynchronous, and a combination of the two. In synchronous updating scheme, the states of the units in a layer are updated as a group prior to propagating the output to the other layer. In asynchronous updating, units in both layers are updated in some order and outputs are propagated to the other layer after each unit update. Lastly, in synchronous-asynchronous updating, there can be subgroups of units in each layer which are updated synchronously while units in each subgroup are updated asynchronously.

Since the BAM also uses the traditional Hebb's learning rule to build the connection weight matrix to store the associated pattern pairs, it too has a severely low memory capacity, The BAM storage capacity for reliable recall was given by Kosko (1988) to be less than minimum (*m*, *n*), i.e., the minimum of the dimensions of the pattern spaces. A more recent study by Tanaka et al (2000) on the relative capacity of the BAM using statistical physics reveals that for a system having n units in each of the two layers, the capacity is around 0.1998*n*.

5.3.11 Discrete BAM

In a discrete BAM, the network propagates an input pattern *X* to the *Y* layer where the units in the Y layer will compute their net input using

$$\text{input } y_j = \sum_{t-1}^{m} x_i w_{ij}$$

And determine their output values using:

$$y_i(t+1) = \begin{cases} +1 & \text{if input } \; y_1 > 0 \\ x_i(t) & \text{if input } \; y_i = 0 \\ -1 & \text{if input } \; y_i < 0 \end{cases} \qquad \text{...(5.3.19)}$$

For $j = 1, 2, \ldots\ldots, n.$

The pattern *y* is propagation base to *x* layer input

$$x_j = \sum_{i=1}^{n} y_i w_{ij}$$

and corresponding equation for $x(t_i)$. Just like the discrete Hopfield model, there is an energy function characterizing the state of the discrete BAM. The energy function due to Kosko (1988) is given by:

$$E = -\sum_{i-1}^{m} \sum_{j-1}^{n} x_i w_{ij} y_j \qquad \text{...(5.3.20)}$$

It is shown by Kosko (1988) that the energy of the BAM decreases or remains the same after each unit update. Therefore, the network will eventually converge to a local minimum that corresponds to a stored associated pattern pair. The energy of the discrete BAM is bounded below by:

$$E = -\sum_{i-1}^{m}\sum_{j-1}^{n} | w_{ij} | \ldots \qquad \ldots(5.3.21)$$

which guarantees that the network will eventually converge to a local minimum corresponding to a stored associated pattern pair.

5.3.12 Continuous BAM

In the continuous BAM, the units use the sigmoid or hyperbolic tangent output function. The units in the *X* layer have an extra external input I_i (not shown in the figure) while the units in the *Y* layer have an extra external input J_j (not shown in the figure) for $i = 1, 2, ..., m$ and $j = 1, 2, ..., n$. These extra external input lead to a modification in the computation of the net input to the units:

$$input\ x_j = \sum_{j-1}^{m} y_j w_{ij} + I_i \qquad \ldots(5.3.22)$$

$$input\ y_j = \sum_{i-1}^{m} y_i w_{ij} + J_j \qquad \ldots(5.3.23)$$

For $i = 1, 2, \ldots m$ and $j = 1, 2, \ldots n$.

The energy function of the Continuous BAM is given as:

$$E = -\sum_{i-1}^{m}\sum_{j-1}^{n} f(x_i)f(y_j)w_{ij} - \sum_{i-1}^{m} f(x_i)I_i - \sum_{j-1}^{n} f(y_j)J_j + \sum_{i-1}^{m}\int_{0}^{x_1} f^1(x_i)x_1 dx_i +$$

$$\sum_{j-1}^{n}\int_{0}^{y_1} f^1(y_j)\, y_j dy_j \qquad \ldots(5.3.24)$$

It is shown by Kosko (1988) that *DE* 0. Therefore, the continuous BAM will eventually converge to a local minimum corresponding to a stored associated pattern pair.

5.3.13 Content Addressable Memory

A content-addressable memory is a type of memory that allows for the recall of data based on the degree of similarity between the input pattern and the patterns stored in memory. It refers to a memory organization in which the memory is accessed by its content as opposed to an explicit address like in the traditional computer memory system. Therefore, this type of memory allows the recall of information based on partial knowledge of its contents.

Suppose we are given a memory of names of several people as shown in the figure below. If the given memory is content-addressable, using the erroneous string "Christopher Columbos" as key is sufficient to retrieve the correct name

"Christopher Colombus." In this sense, this type of memory is robust and fault-tolerant, as this type of memory exhibits some form of error-correction capability.

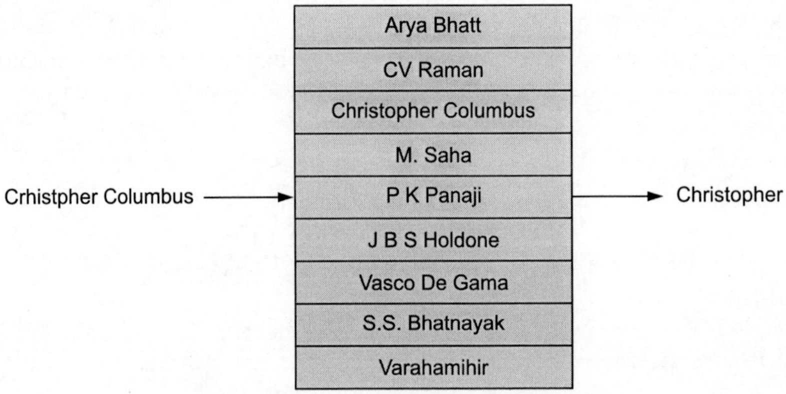

A CONTENT-ADDRESSABLE MEMORY IN ACTION

An associative memory is a content-addressable structure that maps specific input representations to specific output representations. It is a system that "associates" two patterns (X, Y) such that when one is encountered, the other can be recalled. Typically, X_i $\{-1, +1\}m$, Y_l $\{-1, +1\}n$ and m and n are the length of vectors X and Y, respectively. The components of the vectors can be thought as pixels when the two patterns are considered as bitmap images.

5.4 BIDIRECTIONAL ASSOCIATIVE MEMORY (BAM)

BAM is basically a two level non-linear neural network. Kosko extended the unidirectional auto associations to the bidirectional process. The most important property of discrete BAM is its ability to recall stored pairs even in presence of noise.

Features:

1. Taking N training pairs $X_i = (X_{i1}, X_{i2} \ldots X_{in})$ $Y_i = (Y_{i1}, Y_{i2} \ldots Y_{in})$ and allocating X_{ij}, Y_{ij} on / off state the correlation matrix as

$$M = \sum_{t-1}^{n} x_i^T y_i \qquad \text{...(5.4.0)}$$

2. Starting with (α, β) as the initial condition a finite sequence of (α' β') (α^{11} β^{11}) are found such that equilibrium is reached.

$$\beta' = \varphi(\alpha, \beta) \quad \alpha' = (\beta' M^T) \qquad \text{...(5.4.1)}$$

φ stands hard limiting function such that

$$\varphi(5 - 576 - 2) = (1 - 111 - 1)$$

For a given set of pattern pairs (X_i, Y_i) for $i = 1,2 \ldots n$ and their correlation matrix M, a Pair (X^1, Y^1) can be added or an existing pair (X^j, Y^i) can be erased by Special function of the type

$$g_{new} = \begin{cases} 1 & \text{if } f > 0 \\ 0 \text{ or } -1 & \text{if } f < 0 \\ g_{old} & \text{if } f = 0 \end{cases} \qquad \text{...(5.4.2)}$$

where g stands for individual values of new and old and f is the memory equivalent function. When the memory does not affect the result

$$g_{new} = g_{old}.$$

This operation analogous to human memory functions such as learning and forgetting.

5.4.1 Architecture

Let an initializing vector Y be applied at the input (neurons are taken as bipolar binary)

$$x^1 = \Gamma[W_y] \qquad \text{...(5.4.3)}$$

where $\Gamma[.]$ is a non-linear operator defined as the ith output is

$$x_i = \text{sgn}[S_{j=1} w_{ij}\, y_i] \qquad \text{...(5.4.4)}$$

Let us assume that the thresholding operation in the above equation is synchronous and the vector X^1 feeds the layers B of neurons

$$y^1 = \Gamma[W^T a^1]$$

$$y_j^1 = \text{sgn}\left(\sum_{t=1} w_{ij} x_j^1 \right)$$

The sequence of retrieval repeats so that x'' and y'' are computed in terms of recursive update retrieval mechanism

1st FWD PASS $x^1 = \Gamma[wy^0]$ FIRST BWD PASS : $y^2 = \Gamma[w^t x^1]$

2nd FWD PASS $x^3 = \Gamma[wy^2]$

Figure 5.11 shows the simplified diagram of a BAM layers A and B operate in an alternative fashion first transferring the output signal to the right (W) and then towards left (W^T).

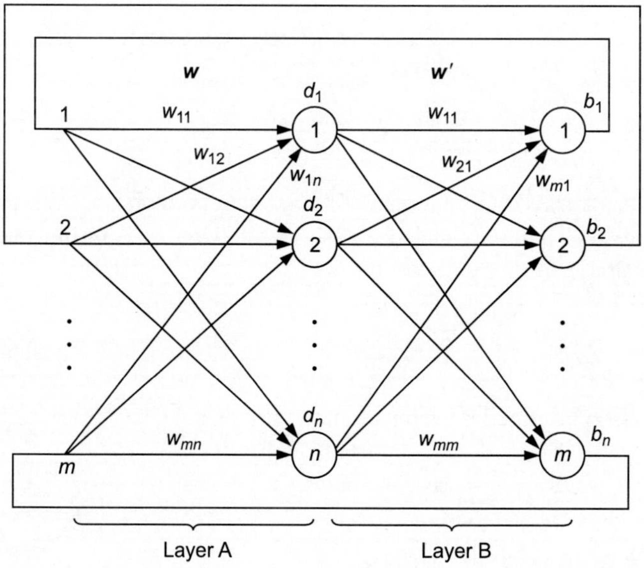

Fig. 5.10

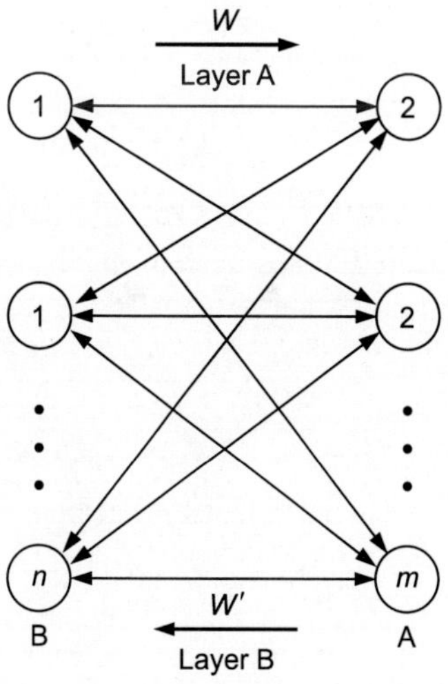

Fig. 5.11

5.4.2 Encoding

Let one of stored pattern $X(m_1)$ be presented.

$$y = \left\| \left[\sum_{m=1}^{p} (y^{(m)} x^{(m)T}) x^{(m^1)} \right] \right\|$$

$$y = \left\| \left[n y^{(m^1)} + \sum_{m \neq m}^{p} y^{(m)} x^{(m)T} x^{(m^1)} \right] \right\| \quad \text{...(5.4.5)}$$

The noise vector in RHS is

$$n = \sum_{m \neq m}^{p} y^{(m)} \left(x^{(m)T} x^{(m^1)} \right) \quad \text{...(5.4.6)}$$

5.4.3 Bidirectional Stability Lyapunov Function

1. A BAM is bidirectionally stable if all inputs converge to a fixed point equilibrium (it provides an example of absolute on global stability)
2. It is a dynamic equilibrium. The same signal information flows back and forth in a bidirectional fixed point.
3. A denotes a binary n-vector in $\{0,1\}n$ and b denotes a m Vector in $\{0,1\}m$ and A enters the BAM system as input then, we can write in what way a BAM equilibrates to the bidirectional fixed point (A_f, B_f) as

$$\begin{aligned} A &\rightarrow \mathrm{P} \rightarrow \mathrm{B} \\ A^T &\rightarrow P^T \rightarrow B^T \\ A^1 &\rightarrow P^T \rightarrow B^1 \\ A^{11} &\rightarrow P^T \rightarrow \mathrm{B}^1 \\ A_f &\rightarrow P \rightarrow B_f \\ A_f &\rightarrow P^T \rightarrow B_f \end{aligned}$$

A^1, A^{11} ..... and B^1, B^{11} ... represent intermediate or transient signal state vectors. A Lyapunov function L maps system state variables to real numbers and decreases with time. In control system Lyapunov function usually involves a quadratic form.

$$\frac{dl}{dt} = \sum_{i}^{n} \frac{\delta L}{\delta x_i} \frac{dx_i}{dt} = \sum \frac{\delta L dx_i}{\delta x_i dt} \quad \text{...(5.4.7)}$$

If I is nXn identity matrix $x = (x_1$ ------ $x_n)$ defined a vector of state variables.

$$L = \frac{1}{2} x I x^T = \frac{1}{2} \sum_{i} xi^2 \quad \text{...(5.4.8)}$$

Taking the example of a dynamical system which describes decay.

$$\frac{dx_i}{dt} = -x_i$$

$$\frac{dL}{dt} = -\sum_i x_i^2 = -\sum_i \frac{dx_i^2}{dt} \qquad ...(5.4.9)$$

In either case $\frac{dl}{dt} < 0$ along the system trajectories as long as at least one variable changes or differs from zero.

At equilibrium $\frac{dl}{dt} = 0$ $\frac{dx_i}{dt} = 0$ for some scalar valued potential function L are stable.

In our case of *A* and *B* $\qquad L = xAx^T \frac{dx}{dt} = xB$

$$\frac{dL}{dt} = xA\dot{x}^T + \dot{x}Ax^T \qquad ...(5.4.10)$$

$$= xAB^T x^T + XBAx^T = x[AB^T + BA]x^T \qquad ...(5.4.11)$$

As *A* represents the identity matrix multiplied by ½. For decay system, *B* represents negative of *I* matrix.

$$-\sum_i \dot{x}_i^2 = x\,[AB^T + BA]\,x^T.$$

A dynamical system is asymptotically stable if and only if the Jacobian of dynamical system has Eigen values with negative real parts. Decay system of $\dot{x}_2 = -x_1$ satisfies this condition as its Jacobian equals the negative of *I* matrix which also accounts for convergence of the linear system.

A Lyapunov function takes care of total system behavior at each moment a single real No. represents the entire system.

This also measures the energy of the physical system Hopfield popularized this neural-network interpretation [A quadratic form in state variables or velocities approximates system energy (potential or kinetic)].

5.5 BAM TRAINING ALGORITHM

5.5.1 BAM Algorithm

For continuous BAM, Kosko proposed there is an inherent capability to transfer the input smoothly and continuously into the respective output in the range of 0–1. This uses a logistic sigmoid function for activation:

For binary input vectors $(s(q), t(q))$, $q = 1, \ldots R$

The weights are determined by

$$w_{ij} = \sum_q (2s_i(q) - 1)\,(2t_j(q) - 1)$$

$$\text{For} \quad f(y_{\text{ing}}) = b_j + \frac{1}{1 + \exp(-y_{\text{ing}})}$$

$$f(x_{\text{ing}}) = b_j + \sum y_j w_{ij}$$

The memory or storage capacity of BAM is min $(n.m)$ where

n = numbers of units in x-layers

m = numbers of units in y-layers

Algorithm

From the process of training we have to attain final weights the input patterns are presented to both x-layers and y-layers the net input and activation are calculated. The signals are sent to X-layers and its net input/activation are found, thus BAM is tested.

1. Weights are intialised to store q vectors (all activities are intialised to 0).
2. For each testing input steps 3-8 are to be followed.
3. Activation of x-layers is set to current input.
4. Input pattern is presented to y-layers.
5. Until activation are conveyed (Steps 6-8 are to be followed).
6. Step 6 Activation unit in Y legs and net input are calculated.
7. Activation of units is done in x-layers, signals are sent to y-layers
8. Convergence test is done.

Stopping iteration is decided by the activation vectors x and y reaching the state of equilibrium.

For discrete BAM

$$S(q) : t(q), q = 1, \ldots q$$

For binary input

$$w_{ij} = \Sigma_q(2s_i\,(q-1)\,(2t_j\,(q-1)$$

For bipolar input vectors, the weight matrix is

$$w_{ij} = \sum_q s_i(q)t_j(q)$$

5.6 ENERGY FUNCTION

For storing a pattern, the value of the energy function for that particular pattern has to occupy a minimum point in the energy level the stability of BAM can be proved by identifying a Lyapunov energy function or E with each state defined by (x, y) in auto associative case.

$E(x) = -XMX^T$ Kosko proposed

$E(X, Y) = -XMY^T$ for hetero association.

When the paired pattern (X, Y) is presented to BAM the neurons change states until a bidirectionally stable state is reached. According to many research papers such a stable state is reached for any matrix M which corresponds to the local minimum of the energy function.

Each cycle of decoding consumes energy. If the energy evaluated by coordinates (X_i, Y_i) as $E = X_iMY_i^T$ does not constitute a local minimum then the point cannot be recalled even though starting point is $\alpha = X_i$ and corresponding $E = -\propto MB^T$.

Example: the patterns give are

0 0 1 1 1 0 0 1	0 1 0 0 0 0 1 1 1
1 1 1 0 0 1 1 1 0	1 0 0 0 0 0 0 0 1
1 1 0 1 1 0 1 0 1	1 0 1 0 0 1 0 1 0

Converting these into bipolar

$X_1 = (-1\ -1\ 1\ 1\ 1\ -1\ -1\ 1)$ $\quad Y_1 = (-1\ 1\ -1\ -1\ -1\ -1\ 1\ 1\ 1\ )$

$X_2 = (\ 1\ 1\ 1\ -1\ -1\ 1\ 1\ 1\ -1\ )$ $\quad Y_2 = (1\ -1\ -1\ -1\ -1\ -1\ -1\ -1\ 1\ )$

$X_3 = (1\ 1\ -1\ 1\ 1\ -1\ 1\ -1\ 1\ )$ $\quad Y_3 = (1\ -1\ -1\ -1\ 1\ -1\ 1\ -1)$

$$\text{The correlation matrix is M } = \begin{bmatrix} -3 & -3 & 1 & -1 & -1 & 1 & -3 & -1 & -1 \\ 3 & -3 & 1 & -1 & -1 & 1 & -3 & -1 & -1 \\ 1 & -1 & -1 & 1 & 1 & -1 & -1 & -3 & 1 \\ -1 & 1 & 1 & -1 & -1 & 1 & 1 & 3 & -1 \\ -1 & 1 & 1 & -1 & -1 & 1 & 1 & 3 & -1 \\ -1 & 1 & -3 & -1 & -1 & -3 & 1 & -1 & 3 \\ 3 & -3 & 1 & -1 & -1 & -1 & 1 & -1 & 3 \\ 1 & -1 & -1 & 1 & 1 & -1 & -1 & -3 & 1 \\ -1 & 1 & 1 & -1 & -1 & 1 & 1 & 3 & -1 \end{bmatrix}$$

Starting $\propto = X_2$

$$\propto M = (13 - 13 - 511 - 5 - 13 - 195)$$

$$\text{Sgn.}(\propto M) = \beta' = (1 - 1 - \ 111 - 1 - 1 - 11)$$

$$\beta' M^T = (5511 - 11 - 115511 - 11)$$

$$\text{Sgn.}(\beta' M^T\ ) = \alpha' (111 \quad 1 \quad 1111 \quad 1)$$

$$\propto' M = (13 - 13 - 511 - 5 - 13 - 195)$$

$$\phi(\propto^1 M) = \beta^{11} = (1 - 1 - 111 - 1 - 1 - 11)$$

$$\beta'' = \beta'$$

So the cycle terminated at $\propto_{eq} = \mu = (-1111\ -1\ -1111\ )$

$$\beta_{eq} = \beta^1 = (1 - 1 - 111 - 1 - 1 - 11)$$

This is a case of recall of incorrect pairs

Computation of energy for (X_2, Y_2) and $(\alpha_{eq}, \beta_{eq})$ given

$$E_2 = -71 \quad E_q = -75$$

$$E_2 = -X_2\ M\ Y_2^T = -(111 - 1 - 1111 - 1)\ MY_2^T$$

Consider Y_2^1 where fifth component -1 of Y_2 has been charged to 1 (one Hamming Distance).

$$E = -X_2 M\ Y_2^{1T} = -73$$

$$-(13-13-511-5-13-195)\begin{bmatrix}1\\-1\\-1\\-1\\-1\\-1\\-1\\-1\\1\end{bmatrix}$$

$$= [\ 13X_1 + (-\ 13X_1) + (-5X-1) + (1X-1) + (1X-1) + (-5X-1) + (-13X-1) + (-19X-1) + (5X_1)]$$

$$= -\ [13 + 13 + 5 - 1 + -1 + 5 + 13 + 19 + 5] = -\ 71$$

As –71 is not lower than –73 it is proved that (X_2, Y_2) is not at its local minimum of E.

It may be noted that BAM cannot guarantee the recall of a training pair or several training pair as the correlation matrix does not ensure that energy of training pair is at local minimum. [A pair can be recalled if and only if it is at local minimum of the energy landscape].

5.6.1 Multi Training Strategy

According the Wang, for recovering a pair using multiple training of order q, matrix M has to be augmented by

$$p = (q-1)X_i^T Y_i \quad (x_i y_i \text{ are in bipolar form}) \qquad \text{...(5.6.1)}$$

The energy function is

$$E(X_i Y_i) = -A_i M B_i^T - (q-1)A_i X_i^T B_i^T \qquad \text{...(5.6.2)}$$

where A_i and B_i are the unipolar forms.

Thus by adjusting q, E_1 can be reduced to low values. This also takes care of energy not exceeding at points one Hamming Distance away from the original location.

5.6.2 Exponential BAM

Exponential BAM (e^{BAM}) has higher capacity for pattern pair storage than conventional BAMs. There is however significant increase in the signal to noise ratio. The energy decreases during recall process ensuring stability. Capacity of e-BAM is increased by the increase in SNR.

If the training pairs are $\{A_1, B_1, A_2, B_2 \ldots A_n B_n\}$ and $X_i\ Y_i$ are the bipolar modes the recall equations are:

$$Y_k = \begin{cases} 1 \text{ if } \Sigma\, Y_{ik}\ b^{x_i x} \geq 0 \\ -1 \text{ if } \Sigma\, Y_{ik}\ b^{x_i x} < 0 \end{cases}$$

$$X_k = \begin{cases} 1 \text{ if } \Sigma\, x_{ik}\ b^{y_i y} \geq 0 \\ -1 \text{ if } \Sigma\, x_{ik}\ b^{y_i y} < 0 \end{cases} \qquad \text{...(5.6.3)}$$

Example:

$X_1 = (-1\ -1\ -1\ -1\ -11\ -1\ -1111\ -11111\ )$ $\quad Y_1 = (1\ -1\ -1)$

$X_2 = (-1\ -1\ -1\ -1\ -1111\ -1\ -111\ -\ 11111)$ $\quad Y_2 = (-11\ -1)$

$X_3 = (-1\ -1\ -1\ -1\ -1\ -1\ 1\ -1\ -111111111)$ $\quad Y_3 = (-1\ -11)$

For retrieving Y_2 against X_2

$$\begin{aligned}(X_2 - X_1) &= (-1 \times -1 + -1 \times -1 + -1 \times -1 + -1 \times -1 \times -1 \times -1 \\ &\quad +1 \times 1 + 1 \times -1 + -1 \times -1 + -1 \times 1 + 1 \times 1 + 1 \times 1 + 1 \times 1 + -1 \\ &\quad + \times -1 + 1 \times 1 + 1 \times 1 + 1 \times 1 + 1 \times 1) \\ &= 1 + 1 + 1 + 1 + 1 + 1 - 1 + 1 - 1 + 1 + 1 + 1 + 1 \\ &\quad + 1 + 1 + 1 = 12.\end{aligned}$$

$$(X_i, Y_i) = 16$$

$$(X_2, Y_2) = 12$$

5.6.3 BAM Energy Function

Consider a physical system of n variable and its Potential energy function E_{xi} measure the displacement from equilibrium of the Ith unit.

$$E = E(X_1 \ldots X_n)$$

By Taylor's equation

$$\begin{aligned}E &= E(0 \ldots 0) + \sum_i \frac{\delta E}{\delta Xi} Xi + \frac{1}{2} \sum_i \sum_j \frac{\delta^2 E}{\delta x_i \delta_{xj}} x_i x_j \\ &= \frac{1}{2} \sum_i \sum_j \frac{\delta^2 E}{\delta x_i \delta_{xj}} x_i x_j \\ &= \frac{1}{2} x A x^T \qquad \ldots(5.6.1)\end{aligned}$$

Same as Lyapunov function $= \frac{1}{2} x A x^T$ where A is symmetric as

$$L = xAx^T$$

$$a_{ij} = \frac{\delta^2 E}{\delta x_i \delta x_j} = \frac{\delta^2 E}{\delta x_j \delta x_i} = a_{ji} \qquad \ldots(5.6.2)$$

If we replace the variable X_i with their time derivatives, the same approx. arg yields the KE of the system as a quadratic form in the velocities gen. $\frac{1}{2} MV^2$ relationships. Negative value of system energy provides a good possibility of Lyapunov function.

For Neural Network; a Lyapunov function need only decrease and be bounded/quadratic equation can grow without bound as the state variables increase in magnitude Neural Network avoid them by using the bounded non - linear signals instead of unbranded neuronal activation.

Stopping point to transients implies Equilibration/point of programmability in Neural Network dynamical systems.

It may correspond to the learning or recall of pattern information on to the solution of combinational optimization problem.

5.7 THEORY OF BAM

As we know, BAM is a hetero associative content addressable memory consisting of two layers. It uses the forward information flow to produce an associative search for the stored association.

Let there be P vector association pairs:

$$\{(X^1, Y^1), X^2\ Y^2 \ \ldots \ \ldots \ X^p, Y^p\}$$

The network becomes a stable state of two pattern reverberation (each pattern at the output of one layers) when the memory neurons are activated the stability pertains to the case of local energy minimum.

The memory processes information in time and involves bidirectional data flow, it is different from linear associator of course both of these are used to store association pairs. It also varies from the recurrent auto associative memory in its update mode.

5.7.1 Stability Theorem for BAM

BAM Stability

If $x^k \rightarrow y^{k+1} \rightarrow x^{k+2}$ and $x^{k+2} = x^k$ then at k^{th} step the memory is said to be bidirectionally stable. This is for the position of the energy function in reading one of its minima

$$E = -\frac{1}{2}x^T Wy - \frac{1}{2}y^T W^T x = -a^T Wy \qquad \text{...(5.7.0)}$$

From the equ. of $x_1^2 = \text{sgn.}\left(\sum_{i=1}^{m} w_{ij} y_i\right) i = 1, 2 \ldots\ldots n$ we have

$$\Delta x_i = \begin{cases} 2 & \sum_{j=1}^{m} w_{ij} y_j > 0 \\ 0 & \sum_{j=1}^{m} w_{ij} y_j = 0 \\ -2 & \sum_{j=1}^{m} w_{ij} y_j < 0 \\ & i = 1, 2 \ldots n \end{cases} \qquad \text{...(5.7.1)}$$

From

$$y_1^1 = \text{sgn}\left(\sum_{j=1}^{m} w_{ij} y_i^1\right) \text{ for } j = 1, 2 \ldots m \qquad \text{...(5.7.2)}$$

$$\Delta y_j^1 = \begin{cases} 2 & \sum_{j=1}^{m} w_{ij} y_j > 0 \\ 0 & \sum_{j=1}^{m} w_{ij} y_j = 0 \\ -2 & \sum_{j=1}^{m} w_{ij} y_j < 0 \\ & j = 1, 2 \ldots n \end{cases} \qquad \text{...(5.7.3)}$$

The gradients are

$$\Delta_x E(x, y) = -Wy$$

$$\Delta_x E(x, y) = -w^T x$$

$$\Delta_{xi}(x, y) = \left(\sum_{j=1}^{m} w_{ij} y_i\right) \Delta x_i, \text{ for } i-1, 2 \ldots n$$

$$\Delta E_{yi}(x, y) = -\left(\sum_{i=1}^{n} w_{ij} y_i\right) \Delta y_i, \text{ for } i-1, 2 \ldots n$$

Compare these expression with update rules loads to $\Delta E \le 0$ the energy changes are non-positive to the energy function is thus a bounded function given by the inequality

$$E(x, y) \ge -\sum_{i=1}^{n} \sum_{j=1}^{m} |w_{ij}|$$

And the memory consoles to a stable point. As no restrictions exist for the choice of the matrix, any arbitrary *nXm* matrix will result in bidirectional stabilities energy is minimum for asynchronous as well as synchronous updates.

5.8 HOPFIELD NETWORK

Architecture shows the basic Hopfield architecture in which all nodes are connected to one another. This network can be used as an associative memory or to solve an optimization problem or as content addressable memory. This is most suitable when exact binary (1 and –1 or 1 and 0) representations are possible as in black and white images with input elements as pixel values.

This network is the second most popular neural network (first being back propagation type). It behaves like child's memory as compared to hetero-associative model based on developed human brain. It can recognize limited number of patterns against exact replica of very similar pattern. It cannot associate one within another different pattern as human adult brain.

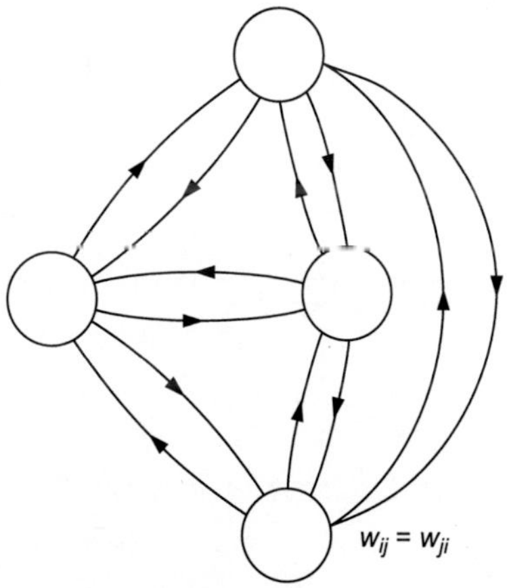

Fig. 5.12

This type of neural networks can be trained to recognize patterns. When presented with a corrupt pattern it can recognize in spite of change in its appearance (Fig. 5.12 shows a pattern corrupted to certain degree).

Basically the network is considered to have a state determined by each individual unit so that when a trained network is presented with a corrupted pattern, its state keeps changing till it reaches a stable steady state. The output of the nodes give the pattern as the network remembers it after the convergence.

The number of patterns which this network can memorise and recall is very much limited. The numbers of classes to be stored should be less than 15% of the number of nodes. If this number is exceeded, the network may have faculty recollection. There are basically two stages of operation:

1. Learning
2. Recognition

5.8.1 Analysis

Consider the basic unit of McCulloch and Pitts neuron model for feedback based. Network in which the output of each unit is fed to all other units with weights W_{ij} for all i and j. We may assume that the output function of each of units is bipolar (±1) so that

$$S_i = F(X_i) = \text{Sgn}(u_i) \quad \text{and}$$

$$U_i = \Sigma_{j=1}^{n} w_{ij} s_i - \theta_i \qquad \text{...(5.8.1)}$$

To start with let $\theta_1 = 0$ the state of one unit depend on states of other units the updating can be done asynchronously or synchronously. In the asynchronous updating, a unit is randomly selected and its new state is determined. Updating using randomly selected units is continued until no further change in the sate take place for all the units. In asynchronous mode, next state is atmost unit Hamming Distance from the current state.

For starting a pattern $u = (u_1 u_2 \ldots\ldots u_n)^T$ in a stable state the updated state becomes same as the current state.

$$\text{sgn}\left[\sum_{j=1}^{N} w_{ij} u_j\right] = u_j \ldots \ldots \text{ for all } i \qquad \text{...(5.8.2)}$$

This is possible if $\quad w_{ij} = \left(\frac{1}{N}\right) u_i u_j$ as

$$\sum_{j=1}^{N} w_{ij} u_j = \frac{1}{N} \sum_{j=1}^{N} u_i u_j u_1 = \frac{u_i}{N} \sum_{j=1}^{N} u_j^2 = u_i \qquad \text{...(5.8.3)}$$

$$as\ u_j^2 = \pm 1 \text{ for the bipolar case}$$

For starting P patterns general Hebbian rule may be applied such that

$$w_{ij} = \frac{1}{N} \sum_{l=1}^{p} u_{li} u_{lj} \qquad \text{...(5.8.4)}$$

The stable U_k will be stable if

$$\text{Sgn}\left[\frac{1}{N} \sum_{j=1}^{N} \sum_{i=1}^{p} u_{li} u_{lj} u_{kj}\right] = u_{Ki} \text{ for all } i \qquad \text{...(5.8.5)}$$

Cancelling the case of $l = K$ and using $u_{kj}^2 = 1$

$$\text{Sgn}\left[u_{kx} + \frac{1}{N} \sum_{j=1}^{N} \sum_{i \neq 1}^{p} u_{li} u_{lj} u_{kj}\right] = u_{Ki} \text{ for all } i \qquad \text{...(5.8.6)}$$

As $U_{ki} = \pm 1$, this is true for all U_{ki} such that the cross term does not change the sign. For $P/N \leq 1$, the cross term is negligible the above regulation is valid for the number of pattern P to be limited to the stored recall storage capacity of network.

5.9 DISCRETE AND CONTINUOUS MODEL

5.9.1 Discrete Hopfield Model

In the discrete Hopfield model, the units use a slightly modified bipolar output function where the states of the units, i.e., the output of the units remain the same if the current state is equal to some threshold value:

$$x_i(t+1) = \begin{cases} +1 & \text{if input } i > \theta_i \\ x_i(t) & \text{if input } i = \theta_i \\ -1 & \text{if input } i < \theta_i \end{cases}$$

For $i = 1, 2, ..., m$ and where t denotes the discrete time.

An interesting property of recurrent type networks is that their state can be described by an energy function. The energy function is used to prove the stability of recurrent type networks. For the discrete Hopfield model with $w_u = 0$ and $w_{ij} - W_{ij}$ using the asynchronous updating scheme, the energy function E according to Hopfield (1982) is defined as:

$$E = -\frac{1}{2}\sum_{i-1}^{m}\sum_{j-1}^{m} x_i w_{ij} x_j - \sum_{i-1}^{m} x_i L_i + \sum_{i-1}^{m} x_i \theta_i$$

where the local minima of the energy function correspond to the energy of the stored patterns. Hopfield (1982) has shown that the energy of the discrete Hopfield model decreases or remains the same after each unit update. Therefore, the network will eventually converge to a local minimum that correspond to a stored a pattern. The stored pattern to which the network converges depends on input pattern and the connection weight matrix.

The energy of the discrete Hopfield model is bounded from below by:

$$E = -\sum_{i-1}^{m}\sum_{j-1}^{m} |w_{ij}| - \sum_{i-1}^{m} |l_i| + \sum_{i-1}^{m} |\theta_i|$$

for all X_k, $k = 1, 2, ..., p$. Since the energy is bounded from below, the network will eventually converge to a local minimum corresponding to a stored pattern.

5.9.2 Continuous Hopfield Model

The continuous Hopfield model is just a generalization of the discrete case. Here, the units use a continuous output function such as the sigmoid or hyperbolic tangent function. In the continuous Hopfield model, each unit has an associated capacitor C_i and resistance r_i that model the capacitance and resistance of real neuron's cell membrane, respectively. Thus the state equation of each unit is now:

$$c_j \frac{d\,\text{input}_j}{dt} = \sum_{i=1}^{m} x_i w_{ij} - \frac{\text{input}_j}{R_j} + I_j$$

where $$\frac{1}{R_j} = \frac{1}{P_j} + \sum_{i-1}^{m} W_{ij}$$

Just like in the discrete case, there is an energy function characterizing the continuous Hopfield model. The energy function due to Hopfield (1984) is given by:

$$E = -\frac{1}{2}\sum_{i-1}^{m}\sum_{j-1}^{m} x_i w_{ij} x_j - \sum_{i-1}^{m} x_i I_i + \sum_{i-1}^{m}\left(\frac{1}{R_i}\right)\iint_0^{x_1} f^{-1}(x)\,dx$$

where f is the output function of the units.

5.10 ALGORITHM FOR STORAGE AND RECALL

Let the network consist of N fully connected units with each unit having hard limit (bipolar) output function the vectors to be stored are U_1 with $1 = 1, 2 \ldots p$ be the vectors to be stored so that $U_{li} = \pm 1, i = 1,2 \ldots N$.

Step 1: The connection weights are assigned values

$$w_{ij} = \frac{1}{N}\sum_{l=1}^{P} u_{li}u_{lj} \; f \text{ or } i \neq 1$$

$$= 0 \text{ if } i = j, \quad 1 \leq i, \quad j \leq N$$

Step 2: The network output is initialized with an input pattern U

$$S_i(i) = U_i \text{ for } I = 1, 2 \ldots N$$

Here $S_i(0)$ is the output of the unit i at time $t = 0$

Step 3: Iteration is continued upto convergence:

$$U_i^{(t+1)} = \left[\sum_{j=1}^{N} w_{ij}s_j(t)\right] f \text{ or } i = 1, 2 \ldots N$$

This is repeated up to the step in which outputs remain unchanged and the outputs at this step represent the stored pattern which best matches the input given.

5.11 STABILITY ANALYSIS

As the updates are continued, the memory comes to equilibrium. In the state of equilibrium denoted by k^{th} step, $a^{k} \to b^{k+1} \to a^{k+2} = a^{k}$. In this position, the memory is said to be bidirectional stable.

Corresponding energy function reaches one of its minima from which decrease is not possible. The energy function taken for such case is $E(a, b) = -a^T w_b$.

Consider the single pattern recall. The summary of threshold bit updates for the outputs of layer A are obtained from:

$$\Delta_i = 1^{a_i}, \ldots n = \begin{cases} 2 \\ 0 \; f \text{ or } \sum_{j=1}^{m} w_i b_j > 0 \text{ or } 0 \text{ or } < 0 \\ 0 \end{cases}$$

Similarly the outputs of B layer are given by $\Delta_j = 1, 2 \ldots m$

The gradients of energy with respect to a and b are $-\overline{w}\overline{b}$ *and* $-\overline{w}^T \overline{a}$

So we may write, $\Delta E_{ai}(\overline{a}, \overline{b}) = -\left(\sum_{i=1}^{m} w_{ij}b_j\right)\Delta a_i$

$$\Delta E_{ai}(\overline{a}, \overline{b}) = -\left(\sum_{i=1}^{n} w_{ij}b_j\right)\Delta b_j$$

These equations imply that $\Delta E \leq 0$. So, the energy changes are non-positive. As the energy E is a bounded function, the memory converges to a stable point. The point under consideration is a local minimum and hence, the memory is bi-directionally stable.

5.12 CAPACITY OF HOPFIELD NETWORK

Hertz et al defined a quantity for capacity as

$$c_i^k = -u_{ki}\frac{1}{N}\sum_{j=1}^{N}\sum_{l\neq k} u_{li}u_{lj}u_{kj}$$

In case c_i^k is negative then the cross term and U_{ki} in eqn...have the same sign so that U_k is stable. In case it is greater than 1, sign of cross term changes and U_k becomes unstable so that the storing cannot be done.

The probability of error is determined by

$$probe = probe\,(c_i^k > 1)$$

The probability of U_{li} equal to ± 1 or –1 is 0.5 c_i^k can be considered a sum of random variables having a binomial distribution with zero mean and variance $\sigma^2 = \frac{P}{N}$ For the case of NP to be large, distribution can be taken as Gaussian with zero mean and δ^2 variance so that

$$Probe = \frac{1}{\sqrt{2\pi\delta}}\int_{1}^{-\delta} e^{-v^2/2\delta^2}\,dv$$

$$= \frac{1}{2}\left[1-\text{erf}\left(\sqrt{\frac{N}{2L}}\right)\right]$$

The values of P_{max}/N – 0.105 and probe = 0.001. So the maximum numbers of patterns which can be stored for a probability of error of 0.001 is

$$P_{max} = 0.105 \text{ N}.$$

In a more comprehensive approach P_{max} is found to be 0.138 N.

5.12.1 Definition of Energy

Assuming the threshold value of unit i and θ_1 the energy function according to hpfield is

$$E = -\frac{1}{2}\sum_i\sum_j {}_j w_{ij}s_is_j + \sum \theta_i s_i \qquad \text{...(5.12.6)}$$

Example: A_3 unit Hopfield network is shown in fig.

Taking the values as

$W_{12} = W_{21} = -0.5$ $\quad$ $W_{23} = W_{32} = 0.4$

$W_{31} = W_{13} = 0.5$ $\quad$ $\theta_1 = -0.1$ $\quad$ $\theta_2 = -0.2$ $\quad$ $\theta_3 = 0.7$

We are supposed to find energy values for each of the state transition table and diagram based states on energy transition table and diagram based states on every changes are to be determined also.

There can be states: 000, 001, 010, 200, 011, 101, 110 and 111

E (0, 0, 0) = 0 as $-\frac{1}{2}\sum w_i s_i s_j = -\frac{1}{2}(w_{11}s_1s_1 = w_{12}s_1s_2 + w_{13}s_1s_3 + w_{21}s_2s_1 + w_{22}s_2s_2 + w_{23}s_2s_3 +$

$$w_{31}s_3s_1 + w_{32}s_3s_2 + w_{33}s_3w_3)$$

And $\Sigma\theta_1 s_i = \theta_1\ s_1 + \theta_2 s_2 + \theta_3 s_3$

$$w_{11} = w_{22} = w_{33} = 0$$

$$E_{(001)} = \theta_3 s_3 = 0.7$$

$$E_{(010)} = \theta_2 s_2 = -\ 0.2$$

$$E(_{100)} = \theta_1 s_1 = -\ 0.1$$

$$E(_{011)} = -\frac{1}{2}(w_{23}s_2s_3 + w_{23}s_3s_2) + \theta_2 s_2 + \theta_3 s_3$$

$$= -\frac{1}{2}(0.4 + 0.4) + (-0.2) + 0.7 = 0.1$$

$$E_{(101)} = -\frac{1}{2}(w_{13}s_1s_3 + w_{31}s_3s_1) + \theta_1 s_1 + \theta_3 s_3$$

$$= -\frac{1}{2}(0.5 + 0.5) - 0.1 + 0.7 = 0.1$$

$$E_{(110)} = -\frac{1}{2}(w_{12}s_1s_2 + w_{21}s_2s_1) + \theta_1 s_1 + \theta_2 s_2$$

$$= -\frac{1}{2}(-0.5 - 0.5) - 0.1 - 0.2 = 0.2$$

$$E_{(111)} = -\frac{1}{2}(w_{12}s_1s_2 + w_{13}s_1s_3 + w_{21}s_2s_1 + w_{23}s_2s_3 + w_{31}s_3s_1 + w_{32}s_3s_2) +$$

$$\theta_1 s_1 + \theta_1 s_2 + \theta_3 s_3$$

$$= -\frac{1}{2}(-0.5 + 0.5 + (-0.5) + 0.4 + 0.5 + 0.4) - 0.1 - 0.2 + 0.7$$

$$= (-0.4) - 0.1 - 0.2 + 0.7 = 0$$

We have to compare $\Sigma_j = 1\ w_{2j} s_j$ with θ_2 for unit 2

State Transition Table

	Unit 1	*Unit 2*	*Unit 3*
000	$(w_{11}s_1 + w_{12}s_2 + w_{13}s_3)$ 0 > –0.1 100 Unit 1 changes state	$(w_{21}s_1 + w_{22}s_2 + w_{23}s_3)$ 0 > –0.2 010 Unit 2 changes state	$(w_{31}s_1 + w_{32}s_2 + w_{33}s_3)$ 0 < –0.7 000 Unit 3 does not changes
001	0.5 > –0.1 101	0.4 > –0.2 011	0 < 0.7 000
010	–0.5 < –0.1 010	0 > –0.2 010	0.5 > –0.7 010
100	0 > – 0.1 100	–0.5 < 0.2 100	0.5 < 0.7 100
011	0 > –0.1 111	0.4 > – 0.2 011	0.4 < 0.7 010
101	0.5 > –0.1 101	– 0.1 > –0.2 111	0.5 < 0.7 100
110	–0.5 < –0.1 010	–0.5 < –0.2 100	0.9 > 0.7 111
111	0 > –0.1 111	–0.1 > –0.2 111	0.9 > 0.7 111

STATE TRANSITION DIAGRAM

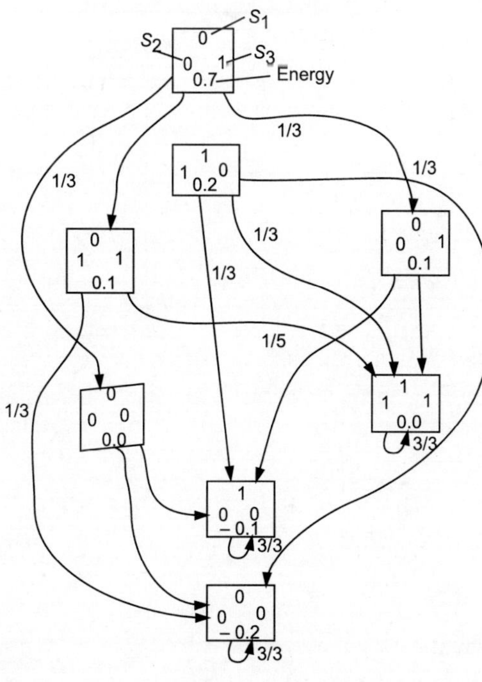

Fig. 5.13

1. Transition is always from a higher energy state to a state with equal or lower energy ($\Delta E \le 0$)
2. Some states 010, 100 and 111 have self-transition probabilities of 1 if these states are reached, the networks remains in these states.
3. Y_3 probability implies that the transition possibility is 1 out 3.

Example: For the Hopfield network shown in figure the weight matrix is

$$W = \frac{1}{3}\begin{bmatrix} 0 & -2 & +2 \\ -2 & 0 & -2 \\ +2 & -2 & 0 \end{bmatrix} \text{ Bias is assumed to be zero}$$

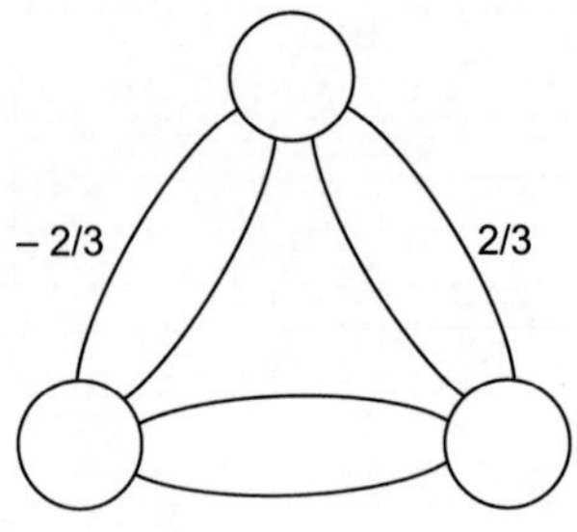

We have to

(*a*) Check the legitimacy of the matrix and

(*b*) Develop the diagram for stable states and flow of the network.

(*c*) $W = \frac{1}{N}(\Sigma x x^t - MI)$ may be checked

(*x* are patterns M = Fund memories)

(*a*) As $W_{ij} = 0$ for all *i* and $W_T = W$ the matrix is legitimate

(*b*) Since there are 3 neurons, we have totally 23 = 8 states

At A (using bipolar Logic) $S_1 = -1$, $S_2 = -1$, $S_3 = -1$

At B $S_1 = 1$ (+ ve in 1 direction), $S_2 = -1$, $S_3 = -1$

At C $S_1 = 1$, $S_2 = -1$, $S_3 = 1$ (+ve in 3 direction)

At D $S_1 = 1$, $S_2 = 1$ $S_3 = 1$

At E $S_1 = -1$, $S_2 = 1$, $S_3 = 1$

At F $S_1 = -1$, $S_2 = 1$ (+ve in 2 direction), $S_3 = -1$

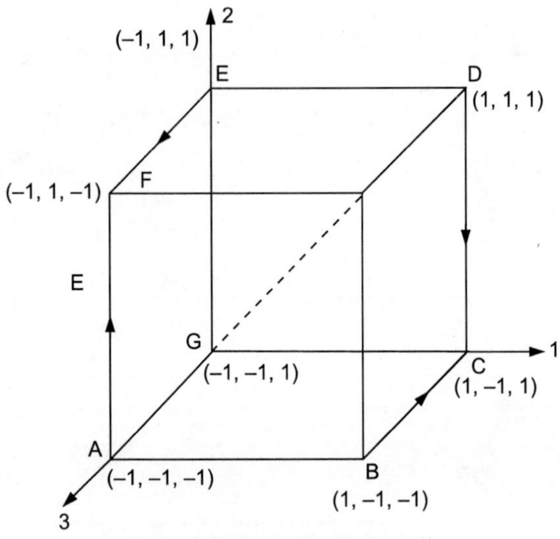

Checking for stability taking C as the first pattern

$$[W]\begin{bmatrix} P \\ A \\ T \\ I \end{bmatrix} = \frac{1}{3}\begin{bmatrix} 0 & -2 & 2 \\ -2 & 2 & -2 \\ 2 & -2 & 0 \end{bmatrix}\begin{bmatrix} 1 \\ -1 \\ +1 \end{bmatrix}$$

$$= \frac{1}{3}\begin{bmatrix} (-2)(-1) & + & 2x_1 \\ -2x_1 & + & -2X_1 \\ 2X_1 & + & -2X-1 \end{bmatrix} = \frac{1}{3}\begin{bmatrix} 4 \\ -4 \\ 4 \end{bmatrix}$$

$$\text{sgn.}\left\{[W]\begin{bmatrix} P \\ A \\ T \\ I \end{bmatrix}\right\} = \begin{Bmatrix} 1 \\ -1 \\ 1 \end{Bmatrix} \textit{ Pattern } 1 = \textit{stability condition is satisified.}$$

By trial it may be seen that for the pattern (–1,1, –1) is also a stable pattern (F)

Flow map is shown by the arrows, this shown symmetry with respect to two stable states of the network. (This symmetry is the result of leaving a neuron in its previous state if the induced local field on it is exactly zero).

If the network is in (1, 1, 1), (–1, –1, 1) or (1, –1, –1) it will converge to (1, –1, 1) after one iteration.

Similarly, if the initial state is (–1, –1, –1) , (–1, 11) or (1, 1, –1) it will converge to second stable state. The gap is just one Hamming Distance.

(c)No. of fundamental memories = 2

$$\frac{1}{3}\begin{bmatrix} 1 \\ -1 \\ 1 \end{bmatrix}[1 \quad -1 \quad 1] + \frac{1}{3}\begin{bmatrix} -1 \\ +1 \\ -1 \end{bmatrix}[-1 \quad 1 \quad -1] - \frac{2}{3}\begin{bmatrix} 1 & 0 & 0 \\ 0 & 1 & 0 \\ 0 & 0 & 1 \end{bmatrix}$$

$$= \frac{1}{3}\begin{bmatrix} 0 & -2 & +2 \\ -2 & 0 & -2 \\ +2 & -2 & 0 \end{bmatrix}$$

5.12.2 Case of Continuous Hopfield Model

$$E = -\frac{1}{2}\sum_{j=1}^{N}\sum_{j=1}^{N} w_{ji}x_i x_j + \sum_{j=1}^{N}\frac{1}{a_j R_j}\int_0^{x_j}\varphi^{-1}(x)dx \qquad \text{...(5.12.7)}$$

$$\text{Intergal } \int_0^{x_j}\varphi^{-1}(x)\,dx \qquad x_i = \sum_j w_{ij}s_j \qquad \text{...(5.12.8)}$$

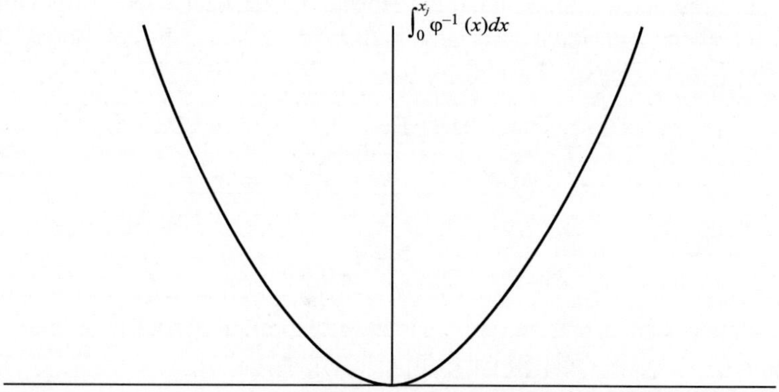

Fig. 5.14

$$f(\lambda x) = \frac{1-e^{-\lambda x}}{1+e^{-\lambda x}} \quad \text{...(5.12.9)}$$

$$T_i \dot{x}_2 = -x_i \sum_i \frac{w_{ij} s_j}{T} \quad [\ T \text{ is time constant}]$$

$$V = -\frac{1}{2}\sum_i \sum_j w_{ij} s_i s_j + \sum_i \int_0^{s_i} f^{(-1)}(s) ds \quad \text{...(5.12.10)}$$

$$\frac{dV}{dt} = -\frac{1}{2}\sum\sum w_{ij} \frac{ds_i}{dt} s_j - \frac{1}{2}\sum\sum w_{ij} s_i \frac{ds_j}{dt} + \sum_i f^{-1}(s_i)\frac{ds}{dt} \quad \text{...(5.12.11)}$$

$$\frac{dV}{dt} = -\sum \frac{ds_i}{dt}\left[\sum w_{ij} s_j - f^{-1}(s_i)\right]$$

$$= \sum \frac{ds_i}{dt}\left[\sum w_{ij} s_j - x_i\right]$$

$$\frac{dV}{dt} = -\sum T_i \frac{ds_i}{dt}\frac{dx_i}{dt} = -\sum{}_i T_i f(x_i)\left(\frac{dx_i}{dt}\right)^2 \quad \text{...(5.12.12)}$$

Since $F(x)$ is on increase F_n $F(x) > 0 = \frac{dv}{dt} \le 0$

Difference between Continuous model and Discrete model:

1. Discrete $\dot{x}_i = 0$ and $v(X) = 0$

   V is called the Lyapunov function of the dynamical system $X >> 1$ Energy function. Error function approaches that of discrete state $\lambda \to \Delta\ \infty$ output is bipolar hyper cable.

Each unit has an extra external input I_i. This extra input leads to a modification in the computation of the net input to the units:

$$\text{input}_j = \sum_{i-1}^{m} x_1 w_{ij} + I_j \quad \text{...(5.12.13)}$$

$$\text{for } j = 1,2, ..., m$$

Unlike the linear associator, the units in the Hopfield model act as both input and output units. But just like the linear associator, a single associated pattern pair is stored by computing the weight matrix as follows:

$$w_k = X_k^T Y_k$$

where $Y_k = X_k$

$$W = \propto \sum_{k-1}^{p} W_k$$

to store p different associated pattern pairs. Since the Hopfield model is an autoassociative memory model, patterns, rather than associated pattern pairs, are stored in memory.

After encoding, the network can be used for decoding. Decoding in the Hopfield model i is achieved by a collective and recursive relaxation search for a stored pattern given an initial stimulus pattern. Given an input pattern X, decoding is accomplished. By computing the net input to the units and determining the output of those units using the output function to produce the pattern X'. The pattern X' is then fed back to the units as an input pattern to produce the pattern X''. The pattern X'' is again fed back to the units to produce the pattern X'''. The process is repeated until the network stabilizes on a stored pattern where further computations do not change the V output of the units.

If the input pattern X is an incomplete pattern or if it contains some distortions, the stored pattern to which the network stabilizes is typically one that is most similar to X without the distortions. This feature is called pattern completion and is very useful in many image processing applications.

During decoding, there are several schemes that can be used to update the output of the units. The updating schemes are synchronous (or parallel as termed in some literatures), asynchronous (or sequential), or a combination of the two (hybrid). Using the synchronous updating scheme, the output of the units are updated as a group prior to feeding the output back to the network. On the other hand, using the asynchronous updating scheme, the output of the units are updated in some order (e.g., random or sequential) and the output are then fed back to the network after each unit update. Using the hybrid synchronous-asynchronous updating scheme, subgroups of units are updated synchronously while units in each subgroup updated asynchronously. The choice of the updating scheme has an effect on the convergence of the network.

Hopfield (1982) demonstrated that the maximum number of patterns that can be stored in the Hopfield model of m nodes before the error in the retrieved pattern becomes severe is around 0.15 m. The memory capacity of the Hopfield model can be increased as shown by Andrecut (1972).

In spite of this seemingly limited memory capacity of the Hopfield model, several applications can be listed (Chaudhuri, Dai, deMenezes, Garcia, Poli, Soper, Suh, Tsai). Discussions on the application of the Hopfield model to combinational optimization problems can be found in (Siu, Suh).

Various widely available text and technical papers by Hassoun, Hertz (et al), Hopfield, McEliece, Ritter, and Volk can be consulted for a mathematical discussion on the memory capacity of the Hopfield model.

5.13 INSTANCE MEMORY BASED LEARNING

Storing and using specific instance improves the performance of several supervised learning algorithms including those for decision trees, classification rules and distributed networks. Instance based learning algorithms do not have any set of abstractions derived from specific instances. This method extends to the nearest neighbour algorithm. Noise becomes a dominant factor in many cases. Noise tolerant decision tree algorithm may be requried for solving such problems.

The salient points are:

1. Instance based learning simply stores the training examples.
2. Each time a new query instance is encountered its relationship to the previously stored examples is examined in order to assign a target function value for the new instance.
3. Instance based learning includes nearest neighbor and locally weighted repression methods.
4. Case based reasoning is also included.
5. "Lazy" or delay processing type of methods are also adapted.

5.13.1 K-nearest Neighbour Learning

The nearest neighbors of an instance are defined in terms of standard Euclidean distance. Let an arbitrary instance be described by the feature vector.

$$< a_1(x), a_2(x) \ldots \ldots a_n(x)T$$

where $a_r(x)$ stands for the value of the rth attribute of instance X.

The distance between two instance x_i and x_j is defined as

$$d(x_i x_j) = \sqrt{\sum_{r=1}^{n} \left(a_r(x_i) - a_r(x_j)\right)^2}$$

Let discrete valued target function of the form $f(w \cdot v)$ have V as a finite set $\{v_1 \ldots\ldots v_n\}$. The value f returned by the algorithm as its estimate is the most common value among K training examples nearest to x. If $k = 1$ then 1-nearest neighbour assigns to $f(x)$ the value of $F(x_i)$ where x_i is the training instance nearest to x. For larger values of K, the algorithm assign the most common value among k nearest training examples among the k training examples nearest to x,. If we choose $k = 1$, then the 1-NEAREST neighbour algorithm assigns to $f(x)$ the value $f(x)$ where x_i is the training instance nearest to x. For larger values of k, the algorithm assigns the most common value among the k nearest training examples.

Figure illustrates the operation of the k-NEAREST NEIGHBOUR algorithm for the case where the instances are points in a two-dimensional space and where the target function is Boolean valued. The positive and negative training examples are shown by "+" and "–" respectively. A query point x, is shown as well. Note the 1-NEAREST NEIGHBOR algorithm classifies x, as a positive example in this figure, whereas the 5-NEAREST NEIGHBOR algorithm classifies it as a negative example.

What is the nature of the hypothesis space H implicitly considered by the k-NEAREST NEIGHBOUR algorithm. Note the k-NEAREST NEIGHBOUR algorithm never forms an explicit general hypothesis f regarding the target function f. It simply computes the classification of each new query instance as needed. Nevertheless, we can still ask what the implicit general function is, or what classifications would be assigned if we were to hold the training examples constant and query the algorithm with every possible instance in X. The diagram on the right side of Figure below shows the shape of this decision surface induced by 1-NEAREST NEIGHBOUR algorithm over the entire instance space. The decision surface is a combination of convex polyhedra surrounding each of the training examples. For every training example, the polyhedron indicates the set of query points whose classification will be completely determined by that training example. Query points outside the polyhedron are closer to some other training example. This kind of diagram is often called the Voronoi diagram of the set of training examples. The k- NEAREST NEIGHBOUR algorithm is easily adapted to approximating continuous-valued target functions. To accomplish this, we have the algorithm calculate the mean value of the k nearest training examples rather than calculate their most common value. More precisely, to approximate a real-valued target function f:(1)ln + !)l we replace the final line of the above algorithm by the line.

$$\hat{f}(x_q) \leftarrow \frac{\sum_{i=1}^{k} f(x_i)}{k}$$

5.13.2 Training Algorithm

For each training example $(x, f(x))$, add the example to the list training examples.

5.13.3 Classification Algorithm

Given a query instance x_q to be classified

Let $X_1 \ldots X_k$ denote the k instances from training examples that are nearest to x_q return.

$$\hat{f}(x_q) \leftarrow \underset{u \in V}{\text{argmax}} \sum_{i=1}^{k} \delta(u, f(x_i))$$

where $S(a, b) = 1$ if $a = b$ and where $6(a, b) = 0$ otherwise.

Fashion and the process for identifying "neighbouring" instances is elaborated accordingly. Case-based reasoning has been applied to tasks such as storing and reusing past experience at a help desk, reasoning about legal cases by referring to previous cases, and solving complex scheduling problems by reusing relevant portions of previously solved problems. One disadvantage of instance-based approaches is that the cost of classifying new "instances" can be high. This is due to the fact that nearly all computation, takes place at classification time rather than when the training examples are first encountered. Therefore, techniques for efficiently indexing training examples are a significant practical issue in reducing the computation required at query time. A second disadvantage to many instance-based approaches, especially nearest neighbor approaches, is that they typically consider all attributes of the instances when attempting to retrieve similar training examples from memory. If the target concept depends on only a few of the many available attributes, then the instances that are truly most "similar" may well be a large distance apart. In the next section we introduce the *k*-NEAREST NEIGHBOUR learning algorithm, including several variants of this widely used approach. The subsequent section discusses locally weighted regression, a learning method that constructs local approximations to the target function and that can be viewed as a *O* generalization of *k*-NEAREST NEIGHBOUR algorithms. We then describe radial basis function networks, which provide an interesting bridge between instance-based and neural network learning algorithms. The next section discusses case-based reasoning, an instance-based approach that employs symbolic representations and knowledge based inference. This section includes an example application of case-based reasoning to a problem in engineering design. Finally, we discuss the fundamental differences in capabilities that distinguish lazy learning methods discussed in this chapter from eager learning methods discussed in the other chapters of this book.

The most basic instance-based method is the *k*-NEAREST NEIGHBOUR algorithm. This algorithm assumes all instances correspond to points in the *n*-dimensional space 8". The nearest neighbors of an instance are defined in terms of the standard Euclidean distance. More precisely, let an arbitrary instance *x* be described by the feature vector

$$\{a_1(x),\ a_2(x)\ \ldots\ a_n(x)\}$$

where $a_r(x)$ denotes the value of the rth attribute of instance x. Then the distance between two instances x_i and x_j is defined to be $d(x_i, x_j)$, where

$$d(x_i x_j) \equiv \sqrt{\sum_{r=1}^{n} (a_r(x_i) - a_r(x_j))^2}$$

In nearest-neighbour learning the target function may be either discrete-valued or real-valued. Let us first consider learning discrete-valued target functions of the form f: $W + V$, where V is the finite set $\{v_1, \ldots v,\}$. The *k*-NEAREST NEIGHBOUR algorithm for approximating a discrete-valued target function is given. As shown there, the value $f(x,)$ returned by this algorithm as its estimate of $f(x,)$ is just the most common value off.

Set of positive and negative training examples is shown on the left, along with a query instance x, to be classified. The 1-NEAREST NEIGHBOUR algorithm classifies *x*, positive, whereas 5-NEAREST NEIGHBOR classifies it as negative. On the right is the decision surface induced by the 1 -NEAREST NEIGHBOUR algorithm for a typical set of training examples. The convex polygon surrounding each training example indicates the region of instance space closest to that point (i.e., the instances for which the 1-NEAREST NEIGHBOUR algorithm will assign the classification belonging to that training example).

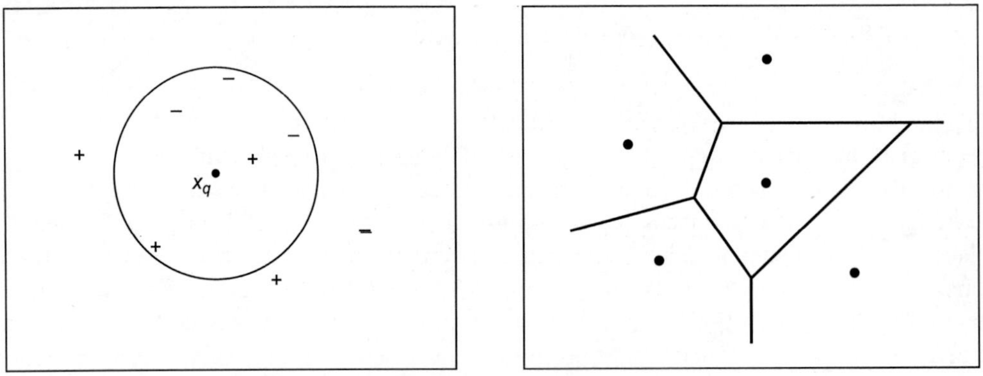

Fig. 5.15 *k-Nearest Neighbor.*

5.13.4 Remarks on k-Nearest Neighbour Algorithm

The distance-weighted k-NEAREST NEIGHBOUR algorithm is a highly effective inductive inference method for many practical problems. It is robust to noisy training data and quite effective when it is provided a sufficiently large set of training data. Note that by taking the weighted average of the k neighbors nearest to the query point, it can smooth out the impact of isolated noisy training examples.

What is the inductive bias of k-NEAREST NEIGHBOUR algorithm? The basis for classifying new query points is easily understood based on the diagrams in the inductive bias corresponds to an assumption that the classification of an instance x will be most similar to the classification of other instances that are nearby in Euclidean distance.

One practical issue in applying k-NEAREST NEIGHBOUR algorithms is that the distance between instances is calculated based on all attributes of the instance (i.e., on all axes in the Euclidean space containing the instances). This lies in contrast to methods such as rule and decision tree learning systems that select only a subset of the instance attributes when forming the hypothesis. To see the effect of this policy, consider applying k-NEAREST NEIGHBOUR algorithm to a problem in which each instance is described by 20 attributes, but where only 2 of these attributes are relevant to determining the classification for the particular target function. In this case, instances that have identical values for the 2 relevant attributes may nevertheless be distant from one another in the 20-dimensional instance space.

As a result, the similarity metric used by k-NEAREST NEIGHBOUR—depending on all 20 attributes-will be misleading. The distance between neighbors will be dominated by the large number of irrelevant attributes. This difficulty, which arises when many irrelevant attributes are present, is sometimes referred to as the curse of dimensionality. Nearest-neighbour approaches are especially sensitive to this problem.

One interesting approach to overcoming this problem is to weight each attribute differently when calculating the distance between two instances. This corresponds to stretching the axes in the Euclidean space, shortening the axes that correspond to less relevant attributes, and lengthening the axes that correspond to more relevant attributes. The amount by which each axis should be stretched can be determined automatically using a cross-validation approach. To see how, first note that we wish to stretch (multiply) the jth axis by some factor = /, where the values $zl \ldots z$, are chosen to minimize the true classification error of the learning algorithm. Second, note that this true error can be estimated using cross validation. Hence, one algorithm is to select a random subset of the available data to use as training examples, then determine the values of $zI \ldots z$, that lead to the minimum error in classifying the remaining examples. By repeating this process multiple times the estimate for these weighting factors can be made more accurate. This process of stretching the axes in order to optimize the performance of *k*-NEAREST NEIGHBOR algorithm provides a mechanism for suppressing the impact of irrelevant attributes.

An even more drastic alternative is to completely eliminate the least relevant attributes from the instance space. This is equivalent to setting some of the z_i scaling factors to zero. Moore and Lee (1994) discuss efficient cross-validation methods for selecting relevant subsets of the attributes for k-NEAREST NEIGHBOR algorithms. In particular, they explore methods based on leave-one-out cross validation, in which the set of m training instances is repeatedly divided into a training set of size m–1 and test set of size 1, in all possible ways. This leave-one out approach is easily implemented in *k*-NEAREST NEIGHBOR algorithms because no additional training effort is required each time the training set is redefined.

Note both of the above approaches can be seen as stretching each axis by some constant factor. Alternatively, we could stretch each axis by a value that varies over the instance space. However, as we increase the number of degrees of freedom available to the algorithm for redefining its distance metric in such a fashion, we also increase the risk of over fitting. Therefore, the approach of locally stretching the axes is much less common.

5.13.5 Hop Fields Content Addressable Memory

The primary function of a content–addressable memory is to retrieve a pattern or item stored in memory in response to the presentation of an incomplete or noisy version of that pattern "An ideal memory could deal with errors and retrieve this reference even from the input "– Warrier (41) .This can also be seen as "a general content–addressable memory would be capable of retiring the entire memory on the basis of sufficient partial information."

The essence of a content addressable memory (CAM) is to map a fundamental memory M_u on to a fixed stable point x_u of a dynamic system. This mapping may be expressed as $M_u \rightleftharpoons X_u$ (left to right is F encoding right to left to decoding)

Hopfield model using the formal neuron of Mc Culloch–Pitts with each neuron having 2 states X_i ± 1 we have 2 Phase:

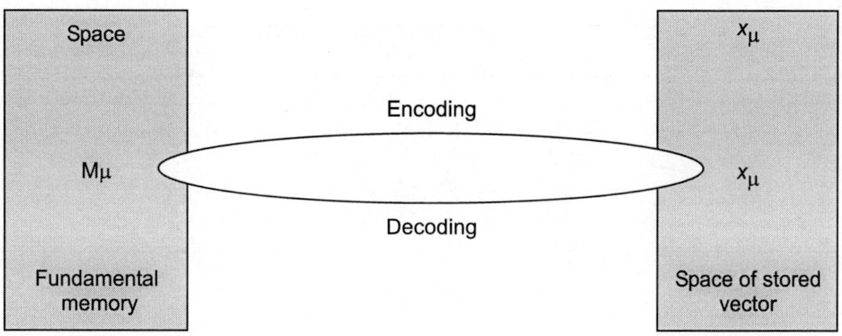

Fig. 5.16

Storage Phase

Let M_{ui} denote the ith element of fundamental memory P_{μ} synaptic weight from neuron i to neuron j.

$$W_{ji} = \frac{1}{N} \sum_{u=1}^{p} M_{uji} M_{u,j}$$

For Hopfiedl $W_{ij} = 0$

so $$W_{ji} = \frac{1}{N} \sum_{u=1}^{p} (M_{\mu} M_{\mu}^{T} - PI)$$

Application of an instance based learning algorithm

- Predicting the stock market index.
- Word sense disambiguation that is based on instance based learning in NLP (Neural learning process).
- Statistical prediction of frequency of highway accidents.
- Statistical data reduction for grid applications.
- Distance criteria in applications such automatic object recognition.
- Anomaly based intrusion detection.
- Web data extraction applications.
- Astronomy (classifying objects – classify pixels in multi- spectral satellite images.
- Automatic estimation of control parameters.
- Medical diagnosis.
- Optical characters quality control in machine vision inspection.
- Symbolic Logic description for applications in design of landscape, building.

SUMMARY

- This unit discusses the application aspects of NN for memory.
- Bilateral associative memory is the main topic.
- Hopfield memory has also been highlighted.
- With many illustrative examples this unit describes the energy analysis.
- The Lapunov theorem in lucid style. Hamming distance calculation has also been shown in a simple manner.

Short Questions

1. What are vectorial rules?
2. What is meant by associative memory?
3. Write about delta rule of pattern association?
4. What is meant b discrete BAM?
5. Define encoding in BAM ?
6. What is meant by multi training strategy?
7. Explain BAM energy function?
8. Define Energy?

9. Explain the difference between continuous and discrete Hopfield model?
10. Write about remarks on K-Nearest Neighbour algorithm?

Long Answer Questions

1. What are static and dynamic associative memories? Explain them with block diagrams?
   Refer 5.1, 5.1.2
2. Explain Briefly about Hamming distance?
   Refer 5.3.5
3. With Neat diagram explain discrete-time bidirectional associative memory. Also explain encoding and decoding information in BAM?
   Refer 5.3.11
4. Explain how Hopfield network can be used to have analog to digital conversion?
5. Explain about Hetero-associative memory and Auto-associative memory?
6. Write about BAM and Derive stability theorem for BAM?
   Refer 5.5
7. Explain about Discrete and continuous Hop field model?
   Refer 5.9
8. Explain about Instance memory based learning?
   Refer 5.13
9. The following vectors need to be stored in a recurrent auto associative memory

   $$S^1 = [1 \quad 1 \quad 1 \quad 1 \quad 1]^T$$

   $$S^2 = [1 \quad -1 \quad 1 \quad 1 \quad -1]^T$$

   $$S^3 = [-1 \quad 1 \quad 1 \quad 1 \quad 1]^T$$

   Compute the weight matrix W.
10. The following Unipolar binary vectors must be stored in the recurrent auto associative memory using the outer product method with the nullification of the diagonal.

    $$S^1 = [1 \quad 0 \quad 0 \quad 1 \quad 0]^T$$

    $$S^2 = [0 \quad 1 \quad 1 \quad 0 \quad -1]^T$$

    $$S^3 = [1 \quad 1 \quad 0 \quad 1 \quad 0]^T$$

    (*i*) Compute matrix W
    (*ii*) Find the analytic expression for the energy function that the memory is minimizing?

❑❑❑

6

FUZZY

6.0 INTRODUCTION

Though Fuzzy Logic has been existing in various usages since time immemorial, it was only in the sixties that Zadeh proposed to separate vagueness as an aspect of uncertainty from randomness. The mathematical base of looking at the tacit vagueness of the natural human language is foundation of Fuzzy Logic. When you tell your friend that you might go to his/her residence soon, you have spoken fuzzily. When we say that the distance is rather long; we imply the length without specifying it. This is also Fuzzy Logic.

Computers fail to understand vague human concepts as the data presented is simple binary Logic. Contrastedly most of human data is not binary. Between black and white there is always a grey area. In our thoughts a person may be good in spite of having some bad quality. A city can be the best and the worst as Dickens thought of London in "Tale of two cities". Basically Fuzzy set theory is Logical, mathematical base necessary to capture our methods of thinking.

Randomness and fuzziness are two different interpretations of uncertainty. Passage of time or additional information can clarify uncertainties of randomness. However the uncertainty in fuzziness is essentially characteristic of words and ideas. According to Heisenberg's Uncertainty Principle, the physical characteristics of a system cannot be simultaneously uneven with unlimited precision. The better we can measure in the position of an electron, the less precisely we know the value of its speed. Thus any attempt to locate electron exactly leads to fuzzification of its moment.

Fuzzy controller with microprocessor was developed by Yama-zaki, Sugeno and others. Combination of Neural Networks with Fuzzy Logic was boosted by Kosko who proposed an adaptive Fuzzy system using Neural Networks. By 1990, Fuzzy Logic became the main stay for vacuum cleaners, cameras, microwave ovens and many other consumer items. The combination of Fuzzy Logic and Neural Networks make electronics even more powerful in the ultimate analysis. This juxtaposition of NN-FL is an epoch comparable with invention of telegraph, usage of vacuum tubes, discovery of transistor operation and development of ICs.

Before examining the properties of Fuzzy Logic sets, it is advisable to review properties, operations and relations of classical sets.

6.1 CLASSICAL SETS

Definition of a Set: A set is basically a collection of objects. The individual objects may be termed as elements or members of the set. "$\in$" stands for "is an element of" where as $\notin$ stands for "is not an element of".

Example: $x \in \chi$: x is an element of set χ.
x is not an element of set χ.

6.1.1 Description of a Set

There are many possible ways to describe or define a particular set

1. A set may be defined by specifying properties of its members.
2. A set may be defined by listing of its members.
3. A formula may be used to describe a set.
4. A set can be defined using an operation on some set.
5. A set may be defined using the membership function.

Example:

1. $A = \{ x | \, x$ is a positive integer $\leq 20\}$
2. $A = \{0, 1, 5, 8, 9, 10, 15, 20\}$
3. $A = \{x_i + 1 = x_i + 2, i = 1, 10 \;\; X_1 = 1\}$
4. $A = \{x | x \in$ BANDC $\}$
5. $\mu_A(x) = 1$ if $x \in A$ and $\mu_A(x) = 0$ if $x \notin A$

6.1.2 Table Shows the Summery of Algebric Rules of Sets

1. If $A \subset B$ and $B \subset A$, $A = B$.
2. If $A \subset B$ and $B \subset C$, then $A \subset C$.
3. $\emptyset \subset A$, $A \subset \emptyset$.
4. $A + B = B + A$, $AB = BA$.
5. $A + (B + C) = (A + B) + C$, $A(BC) = (AB)C$.
6. $A + A = A$, $AA = A$.
7. $A(B + C) = AB + AC$, $A + BC = (A + B)(A + C)$.
8. $A + \emptyset = A$, $A\emptyset = \emptyset$.
9. $A + I = I$, $AI = A$.
10. $A + \bar{A} = I$, $A\bar{A} = \emptyset$
11. $\bar{\emptyset} = I, \bar{I} = \emptyset$
12. $\bar{\bar{A}} = A$, $\overline{A + B} = \bar{A}\bar{B}$ and $\overline{AB} = \bar{A} + \bar{B}$

6.1.3 Cartesian Product of Sets

The name Cartesian is justified as this product leads to the formation of ordered pairs. The Cartesian product of sets A and B is defined as,

$$A \, X \, B = \{(a, b) | a \in A, b \in B\}$$

To show that: $A \, X \, (B \, U \, C) = (A \times B) \;\; U \, (A \, X \, C)$

Let
$$A = \{a_1, a_2, a_3, \ldots\}$$
$$B = \{b_1, b_2, b_3, \ldots\} \text{ and}$$
$$C = \{c_1, c_2, c_3, \ldots\} \text{ then}$$

LHS $= AX(BUC) = \{(a_1, b_1), (a_1, b_2), \ldots (a_1, c_2) \ldots (a_2, b_1), (a_2, c_1), \ldots\}$

RHS $= (AXB)U(AXC) = \{(a_1, b_1), (a_1, b_2), \ldots (a_2, b_1) \ldots . (a_1, c_1) \, (a_1, c_2), \ldots (a_2, c_1), \ldots\}$ = LHS

Selected ordered pairs from the set $A \, X \, B$ form a subset R, termed as the binary relation from A to B. If $A = B$, R is said to a binary relation in A.

6.2 OPERATIONS ON SETS

(a) **INTERSECTION:** $A \cap B = \{X \in A \cap B | \ X \in A \text{ and } X \in B\}$

This has some properties which are similar to those of multiplication of numbers, with universal set I playing the role of unity and empty set Ø playing the part of zero.

$$AB = BA \quad \text{(Commutative law)}$$

$$A(BC) = (AB)C \quad \text{(Associative law)}$$

$$A\emptyset = \emptyset; \ AI = I$$

The unique property (dissimilar to multiplication) is that if BA=B and AA=A

(*b*) **UNION:** $A \cup B = \{x \in A \cup B \,|x \in A \text{ or } x \in B\}$

This has some properties which are similar to addition of numbers.

$$A + B = B + A, \ A\ (B + C) = AB + AC$$

$$A + (B + C) = A + B + C \ A + \emptyset = A \quad A + I = I$$

The unique properties are if $B \subset A$ then $B + A = A$ and $A + A = A$

(*c*) **COMPLEMENT:** The membership function of set A , $\mu_{\bar{A}}(x)$ and $\bar{A}, \mu_{\bar{A}}(x)$ are related by $\mu_{\bar{A}} = 1 - \mu_A(x)$

If $I = \{0, 1\}$ and $A = \{A\}$ then $\bar{A} = \{0\}$

A complement of set A can be defined relative to another set B, rather than I. This is denoted as $\bar{A}_B$.

Venn diagram used to illustrate various operations on the sets. For smoother change over to Fuzzy set theory. The elements x in a set are on horizontal axis, $\mu(x)$ is on vertical axis. For classical (or crisp) sets $\mu(x) = 1$ or 0.

Let $$A = \frac{0.5}{3} + \frac{1}{5} + \frac{0.6}{7} \text{ and}$$

$$B = \frac{1}{3} + \frac{0.6}{5}$$

C is composed of three members 3, 5 and 7 with membership functions $\mu_c(x_1)$, $\mu_c(x_2)$ and $\mu_c(x_3)$

$$\mu_c(x_1) = 0.5 + 1 - 1(0.5) = 1$$

$$\mu_c(x_2) = 1 + 0.6 - 1(0.6) = 1$$

$$\mu_c(x_3) = 0.6 + 0 - 0(0.6) = 0.6$$

$$C = A + B = \frac{1}{3} + \frac{1}{5} + \frac{0.6}{7} = 0.57$$

6.2.1 Bounded Sum

The bounded sum of two Fuzzy sets A and B is denoted by $C = A \oplus B$

Where $\mu_c(x) = \min\ (1, \mu_A(x) + \mu_B(x))$

6.2.2 Bounded Difference

The bounded difference of two Fuzzy sets A and B is denoted by $C = A \otimes B$
Where $\mu_c(x) = \min\{1, \mu_A(x) - \mu_B(x)\}$

6.2.3 Important Relations

$$\mu_A(x) \wedge \mu_B(x) = \mu_B(x) \Theta (\mu_B(x) \Theta \mu_A(x))$$
$$\mu_A(x) \wedge \mu_B(x) = \mu_B(x) + (\mu_B(x) \Theta \mu_A(x))$$

1. Absolute difference: $\mu_A(x) - \mu_A(x) = (\mu_A(x) \Theta \mu_B(x) + (\mu_B(x) \Theta \mu_A(x))$
2. Implication: $\mu_A(x) \rightarrow \mu_B(x) = 1 \Theta (X \Theta Y)$
3. Equivalence: $\mu_A(x) \leftrightarrow \mu_B(x) = 1 \Theta (\mu_A(x) \Theta \mu_B(x)) + (\mu_B(x) \Theta \mu_A(x))$

6.2.4 Convex Combination

The convex combination of n Fuzzy sets $A_1, A_2, A_3, \ldots, A_n$ is a weighted combination of these sets that leads to a Fuzzy set A where

$$\mu_A(x) = w_1(x)\mu_{A1}(x) + \ldots + w_n(x)\mu_{An}(x)$$ with the weights being such that $w_1(x) + \ldots + w_n(x) = 1$

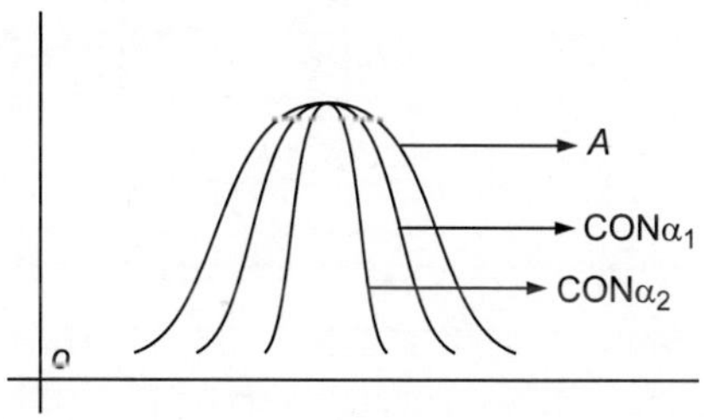

Fig. 6.1 *Effects of CON Operation on the Membership Function.*

CON Operation reduces the values of $\mu_A(x)$ except when $\mu_A(x) = 1$
As $\alpha \rightarrow \infty$, is two valued: 1 or zero
The CON operation makes a set less Fuzzy and leads to crisp sets in the limiting condition.

$$DIL^{\frac{1}{\alpha}}(A) = A^{0.5/\alpha} \text{ Where } \alpha \leq 2.$$

In the limit as $\alpha \rightarrow \infty$ DIL leads to the universe of discourse.

6.2.5 Contrast Intensifier

This operation increases the values of $\mu_A(x)$ which are above 0.5 and reduces those which are below 0.5. The result of applying a contrast intensifier INT to Fuzzy set A is denoted by INT(A)

$$\mu_{INT(A)}(x) \geq \mu_A(x) \quad \text{For} \quad \mu_A(x) \geq 0.5$$
$$\mu_{INT(A)}(x) \leq \mu_A(x) \quad \text{For} \quad \mu_A(x) \leq 0.5$$

This leads to $\mu_{INT(A)}(x) \geq 2\mu_A 2(x)$ for $0 \leq \mu_A(x) \leq 0.5$ ...(6.2.1)

$$\mu_{INT(A)}(x) \le 1 - 2[1 - 2\mu_A(x)]^2 \text{ for } 0.5 \le m_A(x) \le 1 \quad ...(6.2.2)$$

The intensification operation distributes over the union, intersection and product so;

$$\text{INT}(A \cup B) = \text{INT}(A) \cup \text{INT}(B)$$

$$\text{INT}(A \cap B) = \text{INT}(A) \cap \text{INT}(B)$$

$$\text{INT}(AB) = \text{INT}(A)\ \text{INT}(B)$$

6.2.6 Algebraic Sum

The algebraic sum of Fuzzy sets A and B is Fuzzy set C

$$C = A + B$$

where $$\mu_C(x) = \mu_A(x) + \mu_B(x) - \mu_A(x)\ \mu_B(x) \quad ...(6.2.3)$$

This is not to be confused with OR operation.

Let a Fuzzy relation R on $A = \{a_1, a_2, a_3\}$ be

$$R = \frac{0.2}{(a_1,a_1)} + \frac{0.5}{(a_1,a_2)} + \frac{0.4}{(a_1,a_3)} + \frac{1}{(a_2,a_2)} + \frac{0.3}{(a_2,a_3)} + \frac{1}{(a_3,a_2)} + \frac{0.7}{(a_3,a_3)}$$

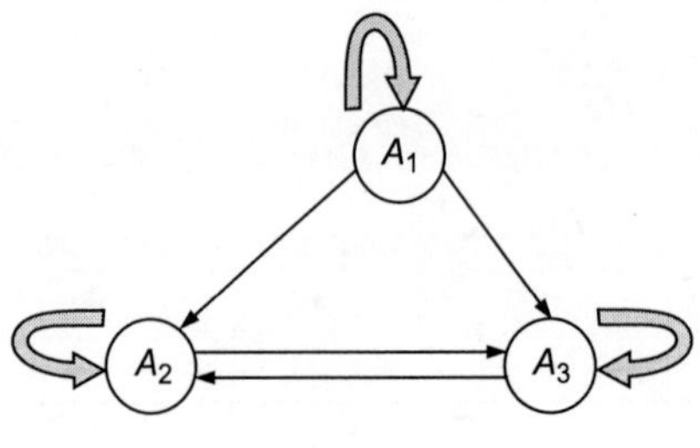

Fig. 6.2

M_R is a 3 × 3 matrix

$$M_R = \begin{array}{c} \\ a_1 \\ a_2 \\ a_3 \end{array} \begin{array}{c} \begin{array}{ccc} a_1 & a_2 & a_3 \end{array} \\ \begin{bmatrix} 0.2 & 0.5 & 0.4 \\ 0.0 & 1 & 0.3 \\ 0.0 & 1 & 0.7 \end{bmatrix} \end{array}$$

Example: Let S and R be binary defined by matrices M_R and M

$$M_S = \begin{bmatrix} 0.7 & 0.8 & 1 \\ 0.2 & 0 & 0 \\ 0.3 & 0 & 0.1 \end{bmatrix} \quad M_R = \begin{bmatrix} 0.1 & 0.6 & 0.3 \\ 0 & 0.1 & 1 \\ 0.1 & 0.7 & 0.8 \end{bmatrix}$$

Then, $R \cap S$ will be described by $M_{R \cap S}$ given by

$$\begin{bmatrix} \min(0.1,0.7) & \min(0.6,0.8) & \min(0.3,1) \\ \min(0,0.2) & \min(0.1,0) & \min(1,0) \\ \min(0.1,0.3) & \min(0.7,0) & \min(0.8,0.1) \end{bmatrix} = \begin{bmatrix} 0.1 & 0.6 & 0.3 \\ 0 & 0 & 0 \\ 0.1 & 0 & 0.1 \end{bmatrix}$$

6.3 FUZZY RELATION

Fuzzy relation from a Fuzzy set A to Fuzzy set B is

$$R = A \times B = \{a, b/a \in A, b \in B\} \qquad ...(6.2.4)$$

When $A = B$, R is known as a Fuzzy relation on A.

Given finite sets A = $\{a_1, a_2, \ldots a_m\}$ $B = \{b_1, b_2, \ldots b_n\}$ a Fuzzy relation $A \times B$ can be expressed as $m \times n$ matrix

$$M_R = \begin{bmatrix} \mu_R(a_1,b_1) & \mu_R(a_1,b_2)\ldots\ldots & \mu_R(a_1,b_n) \\ \mu_R(a_2,b_1) & \mu_R(a_2,b_2)\ldots\ldots & \mu_R(a_2,b_n) \\ \mu_R(a_m,b_1) & \mu_R(a_m,b_2)\ldots\ldots & \mu_R(a_m,b_n) \end{bmatrix}. \qquad ...(6.2.5)$$

This matrix is called Fuzzy matrix. The elements of this matrix have values within the interval [0, 1]

Example: Let $A = \{a_1, a_2, a_3\}$ and $B = \{b_1, b_2, b_3, b_4\}$. Let R be a relation from A to B.

$$R = \frac{0.1}{(a_1,b_3)} + \frac{0.8}{(a_1,b_4)} + \frac{0.8}{(a_2,b_2)} + \frac{0.1}{(a_3,b_1)} + \frac{0.8}{(a_3,b_2)} + \frac{1}{(a_3,b_3)} + \frac{0.8}{(a_3,b_4)}$$

The Fuzzy matrix is a 3 × 4 matrix with entries $\mu_R(a, b)$ that correspond with cell (a, b) such that

$$M_R = \begin{array}{c|cccc} & b_1 & b_2 & b_3 & b_4 \\ a_1 & 0 & 0 & 0.1 & 0.8 \\ a_2 & 0 & 0.8 & 0 & 0 \\ a_3 & 0.1 & 0.8 & 1 & 0.8 \end{array}$$

Fuzzy Graph (Fig. 6.3)

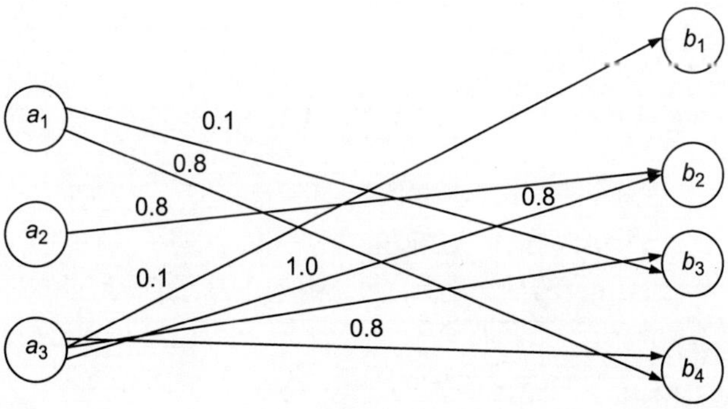

Fig. 6.3

6.3.1 Relation

The concept of relation is usually thought to be a generalization of function use of Cartesian product ensures that many dissimilar sets can be related and ordered pairs can be created.

6.3.2 Basic Definition of Relation

1. A binary relation R, from a set A to set B, is defined as a subset of $A\ X B$. If $(a, b) \in R$ we can say that a is R-related to b $(a\ R\ b)$

2. This can also be expressed by saying that there is a correspondence g from A to B ($g: A \rightarrow B$). A is called the domain and B is called the co-domain.
3. A relation A is any correspondence from A to A.

There are four types of such relations:

(*a*) Reflexive (If $a\ R\ a$ for every $a \in A$).
(*b*) Symmetric (If $a\ R\ b \Rightarrow b\ R\ a$).
(*c*) Anti- symmetric (If $a\ R\ b$ and $b\ R\ a \Rightarrow a = b$).
(*d*) Transitive (If $a\ R\ b$ and $b\ R\ c \Rightarrow a\ R\ c$).

Example: Let A be a set of weather conditions such that A = {snow, rain, sunshine} and let B be a set of items of use such that B = {shoes, boots, coat, umbrella}. The Cartesian product $A \times B$ produces all possible combinations.

$A \times B$ = {(snow, shoes), (snow, boots), (snow, coat), (snow, umbrella), (rain, shoes), (rain, boots), (rain, coat), (rain, umbrella), (sunshine, shoes), (sunshine, boots), (sunshine, coat), (sunshine, umbrella).

A relation R from A to B can be defined by selecting ordered pairs from the set $A \times B$ which satisfy a particular relationship.

For example:

R = {(snow, boots), (snow, coat), (rain, boots), (rain, umbrella), (sunshine, shoes)}.

It is of interest to note that mathematical function is not required to define relation. Such mathematical formation may not exist also. A linguistic description is necessary (rain to umbrella etc).

Example: Let $A = \{1, 2, 3\}$ and $B = \{4, 5, 6\}$ and assume that a correspondence from A to B is defined as "a is half of b" where $a\ A$, $b\ B$ and $(a, b) \in R$.

The Cartesian product $A\ X\ B$ is given by

$$\begin{aligned} A \times B &= \{1, 2, 3\} \times \{4, 5, 6\} \\ &= \{(1, 4), (1, 5), (1, 6), (2, 4), (2, 5), (2, 6), (3, 4), (3, 5), (3, 6)\}. \\ R &= \{(2, 4), (3, 6)\} \end{aligned}$$

We can write $2\ R\ 4$ and $3\ R\ 6$ while $1\ R\ 4$, $1\ R\ 5$, $1\ R\ 6$, $2\ R\ 5$, $2\ R\ 6$, $3\ R\ 4$, $3\ R\ 5$.

Pictorial view:

A = {snow, rain, sunshine}

B = {shoes, boots, coat, umbrella}

R = {(snow, boots), (snow, coat), (rain, boots), (rain, umbrella), (sunshine, shoes)}.

6.3.3 Cartesian Method

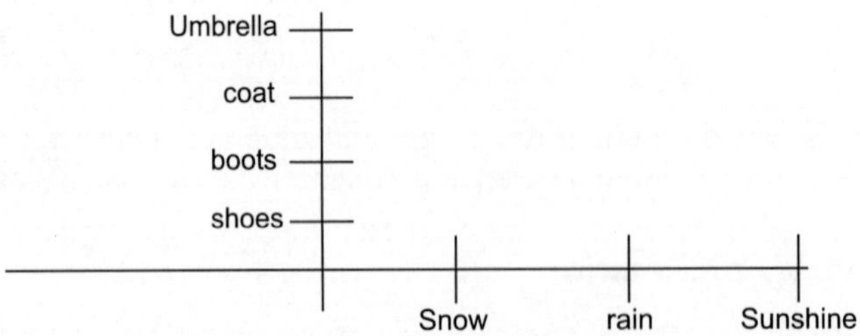

Fig. 6.4

Array method:

umbrella :	0	1	0
coat :	1	0	0
boots	1	1	0
shoes	0	0	1

Matrix form: $\begin{bmatrix} 0 & 1 & 0 \\ 1 & 0 & 0 \\ 1 & 1 & 0 \\ 0 & 0 & 1 \end{bmatrix}$

6.3.4 Array Diagram Method

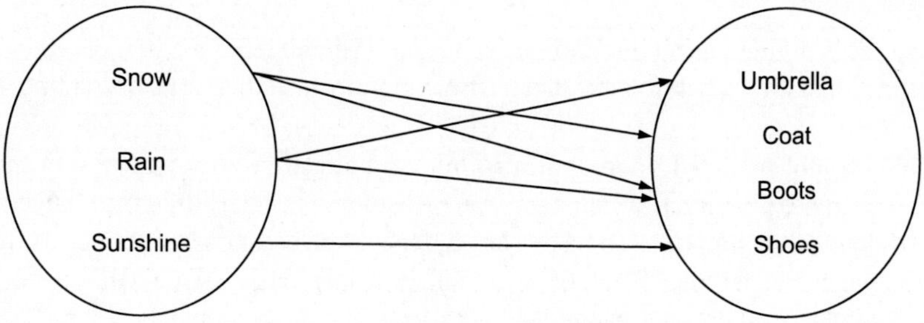

Fig. 6.5

When a relation is from a finite set into itself, there is another method of representation by

Let $A = \{1, 2, 3, 4, 5\}$ and suppose a relation on A is

$$R = \{(1,2), (2, 2), (3, 2), (3, 4), (4, 1), (5, 1), (5, 5)\}$$

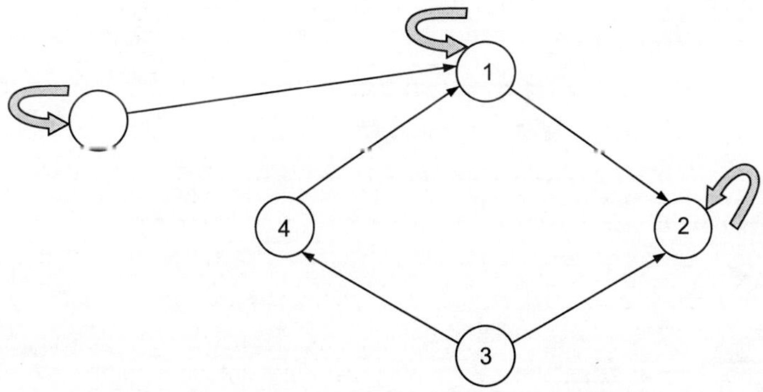

Fig. 6.6

6.3.5 Inverse

If *A* relation *R* goes from domain *A* to codomain *B* the inverse of *R* (R^{-1}) goes from *B* to *A*. The inverse of *R* is defined by

$$R^{-1} = \{(b, a)|(a, b) \in R\} \text{ or } b\, R^{-1}\, a \text{ iff } a\, R\, B.$$

If $A = B$ and $R = R^{-1}$, then *R* is said to be self-inverse, involuntary or symmetric.

If M_R is the matrix of a relation *R*, then the matrix $M_{R^{-1}}$ of R^{-1} is the transpose of matrix *R*.

$$M_{R^{-1}} = M_R{}^T$$

6.3.6 Composition

Let R be a relation from A to B and let S be a relation from B to C. Then R and S lead to a relation from A to C denoted by $R.S$ or RS. This relation is known as the composition of R and S.

$R.S = \{(a, b) \in A \times B \,|\text{there exists } b \text{ in } B \text{ such that } (a, b)\, R \text{ and } (b, c)\, S\}$

Composition R.S can be easily determined by arrow diagram. Matrix of R.S is obtained by multiplying the matrices of R and $S.(M_{R.S} = M_R.M_S)$

6.4 FUZZY SETS

Classical or crisp set is defined as a collection of elements. An element of crisp set either belongs to the set or not. The membership grade or characteristic function is either 0 or 1 leading to two valued Boolean Logic.

Such systems are not natural to perception of human beings. We normally do not and should not take either of two extreme positions. When a universe of students is classified into 'successful' and 'not successful', the students are supposed to have varied degrees of belonging to either of these groups. In successful range we have persons from 60% to 100% (marginally successful to very successful). Among the successful also there is a range. Starting from those who qualify for re-test to those who would be dropped. So the membership to either group is not only one or zero.

Mathematically speaking, a Fuzzy set has a membership function which allows various degrees of membership for the elements of a set under consideration. The membership function may be defined in terms of discrete values or by graph.

If X is a collection of objects denoted by x, then a Fuzzy subset A (symbolically A) (in X is a set of ordered pairs such that $\sim A = \{x, \mu_A(x) | x \in X\}$ where $\mu_A(x)$ is the membership function of x in A. Elments with 0 membership are not listed. In practice, a Fuzzy set will be written as simply A.

Let the universe of discourse be $X = \{1, 2, 3, 4, 5, 6, 7, 8, 9, 10\}$ and

$$A = \{(1, 0.3), (2, 0.5), (3, 1), (5, 0.7), (6, 0.2)\}$$

$$I\,A\,I = 0.3 + 0.5 + 1 + 0.7 + 0.2 = 2.7$$

$$I \times I = 10 \quad II\,A\,II = 2.7/10 = 0.27$$

Graphical representation:

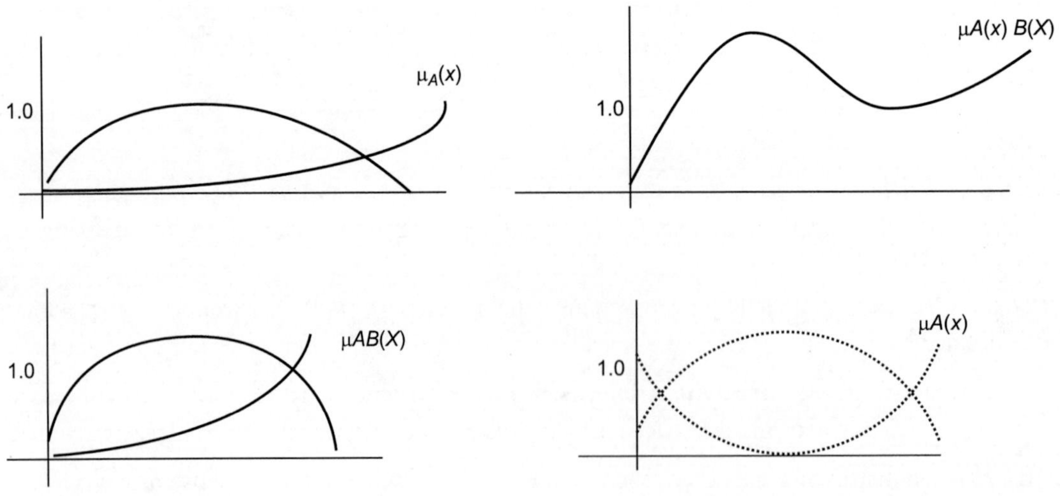

Fig. 6.7

6.4.1 The Support of a Fuzzy Set

The support of a Fuzzy set A is $S(A)$ which is a crisp set of all $x \in X$ such that $\mu_A(x)$ is positive. The element of x in x for which $\mu_A(x) = 0.5$ is called the cross over point. A Fuzzy set whose support is a single element in X for with $\mu_A(x) = 1$ is called a Fuzzy singleton.

THE α-LEVEL SET

The α-level set is a crisp set of elements which belong to the Fuzzy set A atleast to the degree α.

$$A_\alpha = \{x \in X / \mu_A(x) \geq \alpha\}$$

The strong λ-level set or a strong λ-cut is $A^1{}_\alpha = \{x \in X / \mu_A(x) \geq \alpha\}$

A Fuzzy set is said to be convex if all α-level sets are convex

Example: Let

$$A = \{(1, 0.2), (2, 0.5), (4, 0.8), (5, 1), (6, 0.7), (7.0.2)\}$$

$$A_{0.2} = \{1, 2, 4, 5, 6, 7\},\ A_{0.3} = \{2, 4, 5, 6\},$$

$$A_{0.5} = \{2, 4, 5, 6\},\ A_{0.7} = \{4, 5, 6\},\ A_{0.8} = \{4, 5\}, A_1 = \{5\}$$

Let the universe of discourse be $X= \{1, 2, 3, 4, 5, 6, 7, 8, 9, 10\}$ and

$$A = \{(1, 0, 3)(2, 0, 5)\ (3, 1)\ (5, 0, 7)\ (6, 0, 2)\}$$

$$|A| = 0.3 + 0.5 + 1 + 0.7 + 1.2 = 2.7\ |X| = 10$$

$$||A|| = 2.7/10 = 0.27$$

6.5 OPERATION ON FUZZY SETS

6.5.1 Empty and Universal Fuzzy Sets

A fuzzy set is empty iff $u_\phi(x) = 0$ and universal iff $u_x^{(x)} \frac{for\ all}{\forall} x \in x$

Equal sets:

Two fuzzy sets A and B are equal if

$$u_A^{(x)} = u_B^{(x)}\ for\ all\ x \in X$$

6.5.2 Absolute and Relative Complements

The absolute complement (NOT) of a fuzzy set A is denoted by $\overline{A}$ and is given by

$$u_{\overline{A}}^{(x)} = 1 - u_A^{(x)}\ for\ all \in X$$

The relative complement of A with respect to B denoted by $B - A$ is defined by

$$u_{A_R}^{(x)} = u_B^{(x)} - u_B^{(x)} - u_A^{(x)}\ if\ u_B^{(x)} \geq u_A^{(x)}$$

6.5.3 TheIlnion of Fuzzy Sets

The union of two fuzzy sets A and $\boldsymbol{B}$ is a fuzzy set C defined by

$$C = A \cup B \text{ OR } C = A - B$$

Where $u_c^{(x)} = u_A^{(x)} \wedge u_B^{(x)} = \max\left[u_A^{(x)}, u_B^{(x)}\right] x \in X$

6.5.4 The Intersection of Fuzzy Sets

The intersection of two fuzzy sets A and B is a Fuzzy sets is defined by

$$C = A \cap B \text{ OR } C = A - B$$

Where $$u_c(x) = u_A(x) \; A \; u_B(x) = \min[u_A(x), u_B(x)] x \in X$$

6.5.5 Algebraic Product

The product of two fuzzy sets A and B is denoted by AB and is given by

$$AB = \{uA^{(a)} \; uB^{(b)}/x/x \in A, x \in B\} = \{u_A(a) \, u_B(b)/x | \, x \in A, n \in B]$$

Example: Let

$$A = 0.8/3 + 0.9/4 + 0.6/5$$

$$B = 0.7/3 + 0.8/4 + 0.3/5$$

$$AB = 0.56/3 + 0.72/4 + 0.18/5$$

If α is a positive number

$$A^{\alpha} = \{(u_A(x)^a/x/x \in A\}$$

Example: Let

$$A = 0.7/3 + 0.8/5 + 0.3/6 + 1/7$$

$$A^2 = 0.49/3 + 0.64/5 + 0.09/6 + 1/7$$

6.5.6 Concentration and Dilation

These are special cases of A^{α}. The concentration of A is defined as

$$\text{CON}(A) = A^2$$

While the dilation of A is defined as DIL $(A) = A^{0.5}$

Example:

$$A = 0.7/5 + 0.04/7 + 1/9$$

$$\text{DIL}(A) = 0.84/5 + 0.2/7 + 1/9$$

These two operation are of importance in Linguistic Hedges.

6.5.7 Properties of CON and DIL

$\text{CON}(A + B) = \text{CON}(A) + \text{CON}(B)$ $\quad (A + B)^2 = A^2 + AB + BA + B^2$

$\text{CON}(A \; ? \; B) = \text{CON}(A) + \text{CON}(B)$ $\quad AB + BA \;\; C \;\; A^2 + B^2$

$$(A + B)^2 = A^2 + B^2$$

Similarly,

$$\text{CON}(A.B) = \text{CON } A.\text{CON}(B)$$

$$\text{CON}^2(A) = A4 = \text{CON}^{\alpha}(A) = A^{\propto}$$

To show that DeMorgan's laws are valid for fuzzy sets:

$$\overline{A \cup B} = \bar{A} \cap \bar{B} \quad \text{...(6.1)}$$

$$\overline{A \cap B} = \bar{A} \cup \bar{B} \quad \text{...(6.2)}$$

To prove that Eq. (6.1) is true we have to show that

$$\text{Max } [\mu A(x), \mu B(x)] = \text{Min}[(1-\mu A(x)), 1-\mu B(x)]$$

If $\mu A(x) > \mu B(x)$ then both sides are equal to $1-\mu A(x)$. if $\mu A(x) < \mu B(x)$ then both sides are $1-\mu B(X)$.

6.5.8 Models of Fuzzy AND, OR, INVERTER

1. Figure 6.8 shows a circuit which explains the concept fuzzy and the fuse with the lower rating decides the maximum current which can go through the circuit and so, the brightness of light is
   Brightness α min $[A, B]$

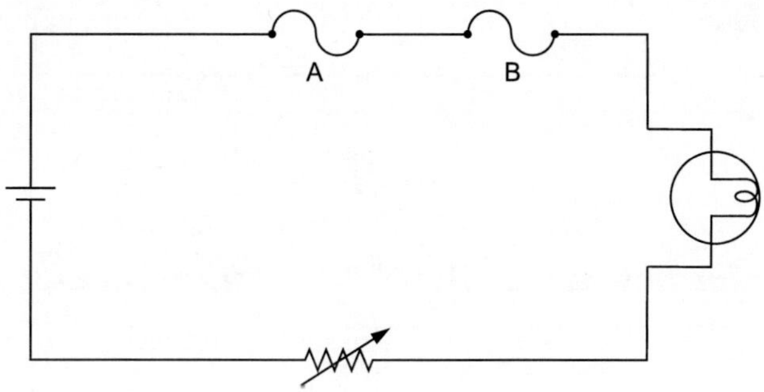

Fig. 6.8

2. Figure shows a circuit for OR: brightness α min $[A, B]$

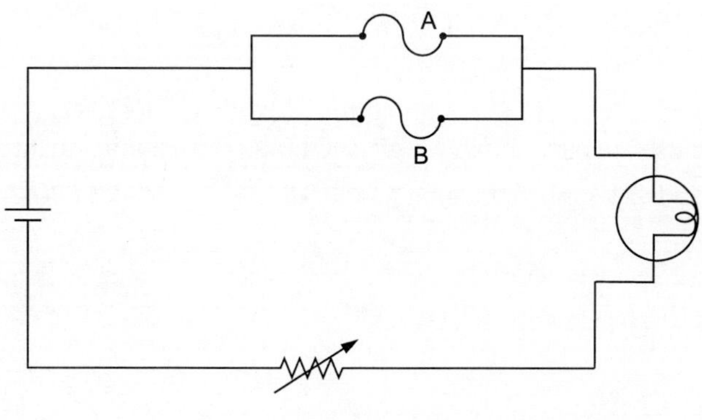

Fig. 6.9

3. Figure shows the operation of Inverters. If the current in the circuit is unity and current *n* the parallel resistor is I then the current through the large is $(1 - I)$ so the brightness is given by Brightness α $[1 - I]$

6.5.9 Algebraic Operations

Cartesian multiplication: The Cartesian product of two fuzzy set A and B is a fuzzy Set C denoted by $A \times B$ and given by

$$C = AXB = \left\{ \frac{\mu\ (x)}{(a, b)} 1 \in A, b \in B \text{ and } U_c^{(c)} = \min\left(\mu_a(a), \mu_b(b)\right) \right\}$$

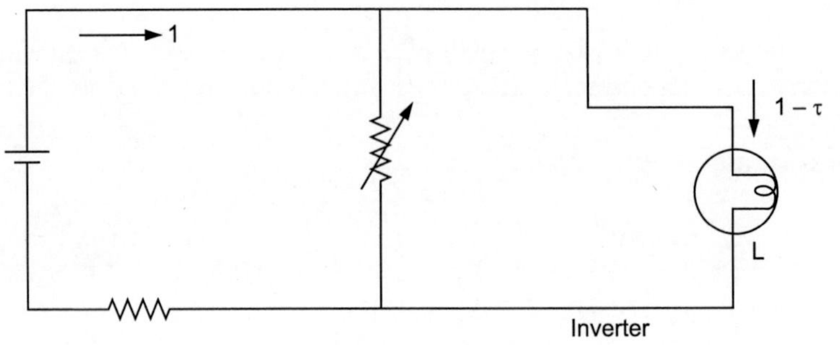

Fig. 6.10

SUMMARY

- This Unit introduces the concepts of fuzzy logic.
- The algebraic relationships are also covered.
- Various types of representation of fuzzy logic have been discussed.
- The difference between fuzzy logic and usual crisp logic can be appreciated by going through this unit.
- In day to day there are so many examples of fuzzy logic based activities.
- In complex application such simple concepts are only effective if we are in a position to understand.
- By seeing a vehicle ahead of us and going the proximity (close, very close) we put the brake. Close and very close are part of fuzzy logic and brake is a definite quantity based on crisp logic.
- This unit in a way is a simple launching pad for delivering deeper into the fuzzy world.

Short Questions

1. What is meant by CON?
2. Define Set?
3. What is meant by bounded sum and bounded subtraction?
4. Define Relation?
5. Explain Cartesian Method?
6. What is Array diagram method?
7. What is meant by concatenation and dilation?
8. Give the properties of CON and DIL?

Long Answer Questions

1. Give and explain properties of crisp sets?
2. Explain following term
   (*i*) CON
   (*ii*) DIL
   (*iii*) Membership function
   (*iv*) CRISP Relation
   Refer 6.5.7
3. Explain operation of Fuzzy sets and Fuzzy relation?
   Refer 6.5 and 6.3
4. Let $\overline{x} = \left\{\frac{1}{2} + \frac{0.5}{3} + \frac{0.3}{4} + \frac{0.2}{5}\right\}$ and $\overline{y} = \left\{\frac{0.5}{2} + \frac{0.7}{3} + \frac{0.2}{4} + \frac{0.4}{5}\right\}$ be two discrete fuzzy sets.
   Calculate several operations on these sets. Assume membership for element in both sets be zero?
5. Explain properties and operations of fuzzy relations?
6. Explain about
   (*a*) Composition of fuzzy relations?
   (*b*) Projection of fuzzy relations
   (*c*) Cylindrical Extensions
   (*d*) Cylindrical Closure

❑❑❑

7

FUZZY LOGIC SYSTEM COMPONENTS

7.0 INTRODUCTION

This unit discusses various aspects of Fuzzy Logic such as membership value assignment, development of rule base decision making system and defuzzification. These are stepping stones to understand application of Fuzzy Logic and conversely design Fuzzy systems. Basically rule based systems are extension of IF_THEN Logic which can be used to make the decision in Logic system used in controls. For example, the brake applied by a driver may be specific, but the decision in doing it is based on IF THEN Logic of Fuzzy sets (of "to near" or "just near" with respect to preceding vehicle.) Any Fuzzy data needs to be made crisp to make it applicable (like the specific pressure in the braking). There are many methods of defuzzification depending on the geometry of representation of data. The results of all these methods are different from each other. Normally the suitability of the methods various depending on the applications.

The concepts developed in this unit will be useful in discussing practical cases taken up in Unit VIII. Unit that time some of the ideas of the unit may look "Fuzzy" to the students!

7.1 FUZZIFICATION

The process of fuzzification transforms a set (Fuzzy or Crisp) to an approximation which is more Fuzzy. This process can be considered to a be generalization of dilation process.

The basic feature of fuzzification is point fuzzification. Point fuzzification transforms a singleton set $1/u$ in u to a Fuzzy set u which varies around u the symbol ~(approximate) is used to denote a fuzzifier. For example, ~70 represent the Fuzzy set of real numbers which are approximately equal to 70 (selected from the universe of discourse).

In order to express the dependence of u on $\sim u \sim u$ is written as $u = K(u)$ the fuzzy set $K(u)$ is called the kernel of Fuzzification.

EXAMPLE 7.1: Let the universe of discourse be

$$U = 1 + 2 + 3 + 4 + 5$$

Solution: Let A be a singleton set given by ½. We can define

$$K(2) = \frac{0.5}{1} + \frac{1}{2} + \frac{0.7}{3} + \frac{0.2}{4} + \frac{0.1}{5}$$

$$\sim A = \left\{\frac{0.5}{1}, \frac{1}{2}, \frac{0.7}{3}, \frac{0.2}{4}, \frac{0.1}{5}\right\}$$

The choice of the process of fuzzification (description of $K(u)$) depends on the meaning of the sets and criteria.

Considering a general case of $A = u_1/x_1 \pm \ldots \frac{u_n}{x_n}$. We may observe that is every member and its membership grade are independent of other member and their membership grades, we can assume that the process of fuzzification is linear and super position may be applied (fuzzifying each member alone)

The process fuzzification of $\frac{u_i}{x_i}$ can be two types:

(*a*) Where u_i is kept constant and x_i is fuzzified (support fuzzification)

(*b*) u_i is fuzzified and x_i is kept constant (grade fuzzification)

7.1.1 Support Fuzzification (s-Fuzzification) and Grade Fuzzification (G-Fuzzification)

Support fuzzification of A is denoted by $SF(A; K)$ and is defined as

$$A = Sf(A;K) = u_i\, K(x_i) + \ldots + u_n\, K(x_n)$$

Here, $u_i\, K(x_i)$ is a Fuzzy set which is the product of scalar constant u_i and a Fuzzy set $K(x_i)$.

EXAMPLE 7.2: Let the universe of discourse be

$$\mathbf{U = 1 + 2 + 3 + 4 \text{ and let}}$$

$$\mathbf{A = \frac{0.8}{1} + \frac{0.5}{2}}$$

Solution: Assume

$$K(1) = \frac{1}{1} + \frac{0.3}{2}$$

$$K(2) = \frac{1}{2} + \frac{0.3}{1} + \frac{0.2}{3}$$

$$\text{SF}(A; K) = 0.8\left(\frac{1}{1} + \frac{0.3}{2}\right) + 0.5\left(\frac{1}{2} + \frac{0.3}{1} + \frac{0.2}{3}\right)$$

$$= \frac{0.8}{1} + \frac{0.24}{2} + \frac{0.5}{2} + \frac{0.15}{1} + \frac{0.1}{3}$$

Since the union is for maximising

$$\text{SF}(A; K) = \frac{0.8}{1} + \frac{0.5}{2} + \frac{0.1}{3}$$

A noteworthy point is that a dilation of crisp set yields the same crisp set which is not true in general for S-fuzzification.

7.1.2 Grade Fuzzification (g-Fuzzification)

Grade fuzzification A is denoted by $GF(A;K)$ and defined as

$$A = GF(A;K) = K(u_i)/X_1 + \ldots K(u_n)/x_n$$

Where $K(u_i)$ is the point fuzzification of u_i.

EXAMPLE 7.3: Let the universe of discourse be $U = 1 + 2 + 3 + 4$ and let $A = \frac{0.8}{1} + \frac{0.5}{2}$

$$GF(A;K) = A = \frac{0.8}{1} + \frac{0.5}{2}$$

We define $K(0.8) = \frac{1}{0.8} + \frac{0.7}{0.6} + \frac{0.3}{0.5}\ K(0.5) = \frac{1}{0.5} + \frac{0.6}{0.4} + \frac{0.5}{0.3}$

$$A = \frac{\frac{1}{0.8} + \frac{0.7}{0.6} + \frac{0.3}{0.5}}{1} + \frac{\left(\frac{1}{0.5} + \frac{0.6}{0.4} + \frac{0.5}{0.3}\right)}{2}$$

7.2 MEMBERSHIP FUNCTION

A Fuzzy set is basically a set with smooth unsharp boundary. It generalizes the notion of membership from a black and white catergorisation into one that allows partial membership. The idea of membership in Fuzzy sets becomes a matter of degree, (which is bounded by 0 and 1) mathematically, a Fuzzy set is characterized by a mapping from its universe of discourse into the interval [0,1]. This mapping is the membership function of the set Fuzzy set. For example, high temperature may be defined using the membership function shown in Fig. 7.1.

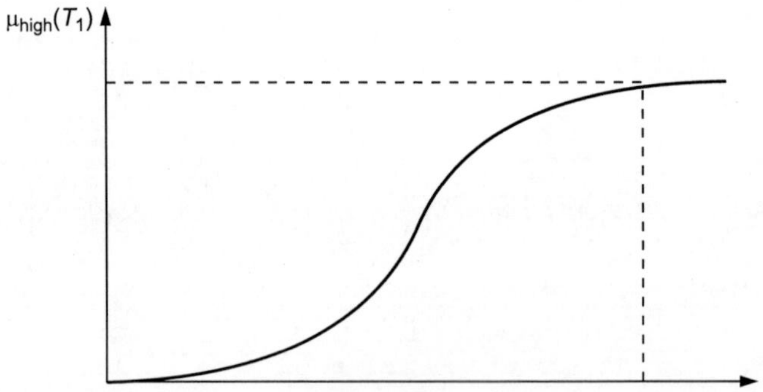

Fig. 7.1

A set of core Fuzzy sets about a variable forms a term set which is the basis of generating composite Fuzzy sets using hedges. For example, low/medium/high whose membership are shown in Fig. 7.2 form a complete term set for the temperature T of a process Term set is said to be complete when for any given element in the universe of discourse there is one term in the term set that covers the given element.

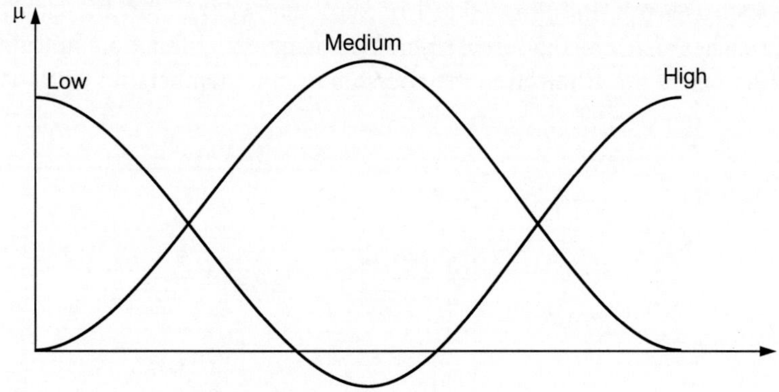

Fig. 7.2

A Fuzzy set may be defined in two ways.

1. By enumerating membership values of these elements in the set (partially or completely)
2. By mathematically defining membership function. Generally a Fuzzy set can be defined through enumeration using the expression.

$$A = \sum_i u(x_i)/x_i$$

7.2.1 Types of Membership Function

(*a*) Triangular

$$\text{Triangle }(x: a, b, c) = \begin{cases} 0; & x < a \\ x - a/b - a; & a \le x \le b \\ c - x/c - b; & b \le x \le c \\ 0 & x > x \end{cases} \quad ...(7.2.1)$$

(*b*) Trapezoidal

$$\text{Trapezoid }(x: z, b, c, d) = \begin{cases} 0; & x < a \\ (x - a)/b - a; & a \le x \le b \\ 1 & b \le x \le c \\ d - x/d - c & c \le x < d \\ 0 & x \le d \end{cases} \quad ...(7.2.2)$$

(*c*) Gaussian

$$\text{Gaussian }(X; m\sigma) = \exp\left[\frac{(x - m)^2}{\sigma^2}\right] \quad ...(7.2.3)$$

(*d*) Bellshpaed

$$\text{bell }(x: a.b.c) = \frac{1}{x + \left|\frac{x - c}{a}\right|^{2b}} \quad ...(7.2.4)$$

Sigmoidal

$$\text{Sgn }(x: a, c) = \frac{1}{1 + e^{-a(x - c)}} \quad ...(7.2.5)$$

As parameters increases transition from 0 to 1 become sharper. The function approximates the step function as '∝' approaches ∞. $\mu(c) = 0.5$ for all sigmodial membership function indicating the breakeven point of the definition where transition from membership to non membership takes place.

S TYPE

$$S(x: a, b) = \begin{cases} 0; & x < a \\ 2\left(\dfrac{x-a}{b-a}\right)^2; & a \le x \le \dfrac{a+b}{2} \\ -2\left(\dfrac{x-a}{b-a}\right)^2; & \dfrac{a+b}{2} \le x < b \\ 1 & x \ge d \end{cases} \qquad ...(7.2.6)$$

π Membership

$$\pi_1(x; a, b) = \frac{1}{1+\left(\dfrac{x-a}{b}\right)^2} \qquad ...(7.2.7)$$

π_1 function is used for good stopping accuracy

$$\pi_2(x, 1w, 1p, rp, rw) = \begin{cases} \dfrac{lw}{lp + lw - x} & x < lp \\ & lp \le x \le rp \\ \dfrac{rw}{x - rp + rw} & x > rp \end{cases}$$

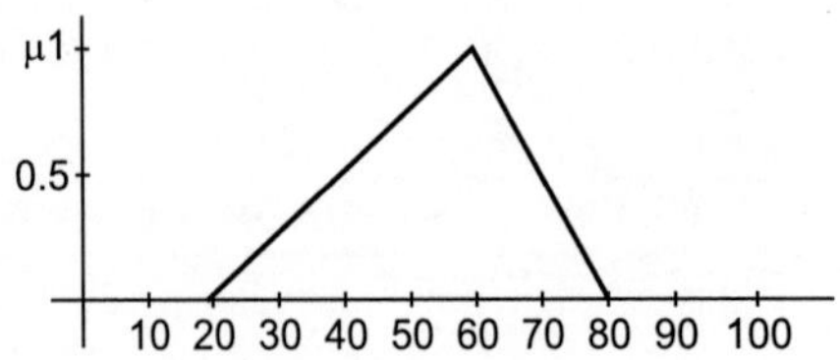

$(x : 20, 60, 80)$ Triangular

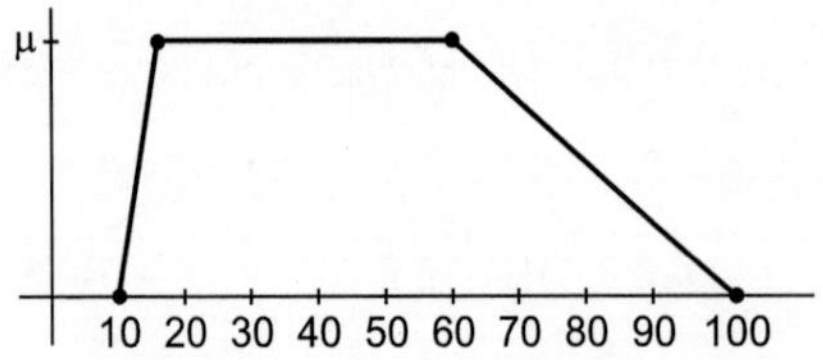

$(x : 10, 20, 60, 95)$ Trapezial

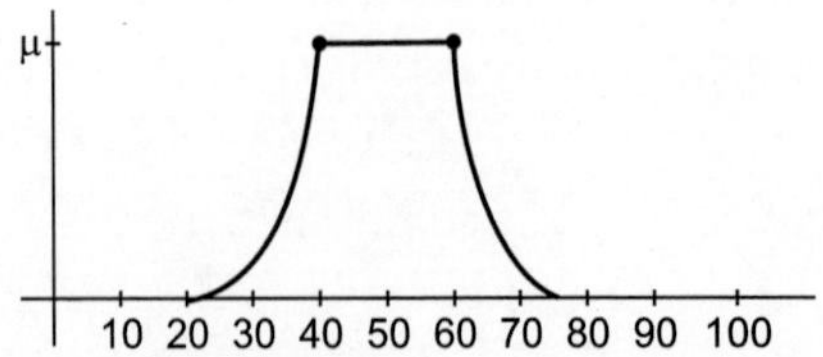

$(x : 20, 4, 50)$ Bell

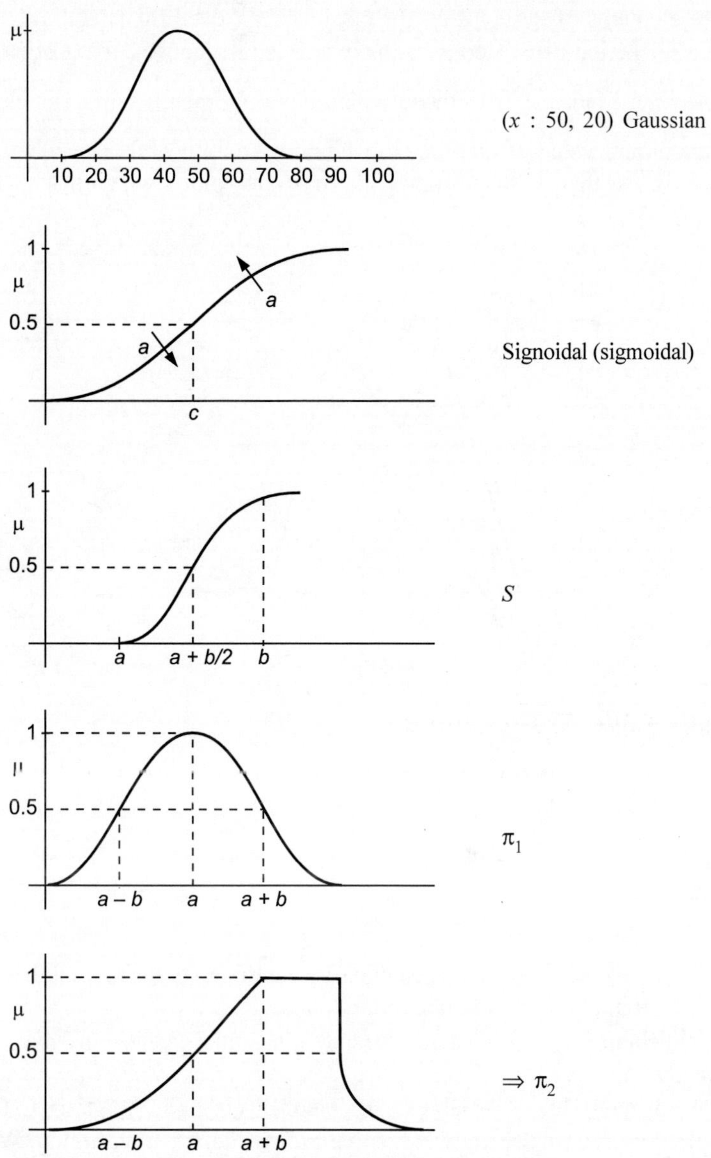

Fig. 7.3

7.2.2 Features of Membership Function

Core: It is the portion of the domain containing elements which have full membership

The set $\{x \in X\, u_A(x) = 1\}$ is the core

Boundary: It is the portion of the domain containing elements which have non-zero membership (but not complete membership). $\{x \in X, 0 < u_A\{x) < 1\}$ give the boundaries.

Support: It is the portion of the domain which has elements of non-zero membership the set $\{x \in X, u_A(x) > 0\}$ gives the support.

Convex: If x, y, z are elements of A such that for $x < y < z$ $u_A(y) > \min\{u_A(x), u_A(z)\}$ then the set is convex fuzzy set the intersection of two convex fuzzy sets is also convex fuzzy set.

Cross over: The cross over points refer to those which have 0.5 membership values.

Height: This is the maximum value of membership function (max $u_A(x)$). The fuzzy set is said to be subnormal if the height is less then 1 it is normal if there is one point x which has $u_A(x) = 1$.

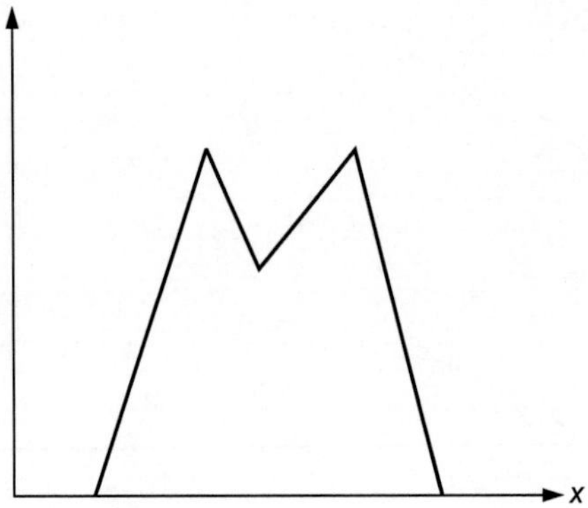

Fig. 7.4

7.2.3 Membership Value Assignments

The membership values can be assigned by the following methods

1. Intuition.
2. Inference
3. Rank ordering
4. Inductive reasoning

1. Intuition: It is derived from the capacity of humans to develop membership functions S through their own intelligence and understanding.
It involves some knowledge about an issue contextual and it can also involve truth values based on linguistic knowledge.
For example, for temp, fuzzy variables can be cold, very cold, hot, very hot.

2. Inference: In this case we use knowledge to perform deductive reasoning. We wish to deduce or infer a conclusion given a body of facts and knowledge.
For example, in evaluating the nature of Δ^{le} we can say

(*a*) Nearly, isosceles Δ^{le}
(*b*) Nearly right angled Δ^{le}
(*c*) Nearly equilateral Δ^{le}

3. Rank ordering: In this case preferences of assessment by a single individual, or a poll or other opinion based methods can be used to assign membership values to fuzzy variables.
Angular fuzzy sets: These differ form ordinary fuzzy sets only in their co-ordinate description. These are defined only in universe of analysis form 0 to 2π.
For example, AC motors has a special advantage over DC Motors in the rotation of armatures (DC Motors rotate unidirectional, AC motors can rotate in both directions)

In case of stepper motors, we have variable reluctance and a permanent magnet In case of variable type, stators have a common form and rotors have a common shaft. Stators are pulse excited and rotors are unexcited. Variable reductance are used in incremental motion control systems.

The directional rotation of variable type can be regulated by the kind of pulse applied, by defining the following terms related to motion of motor. We can form an angular fuzzy set

1. Completely anti clock wise $\theta = \pi/2$.
2. Partially anti clock wise $\theta = \pi/4$.
3. No motion with motor $\theta = 0$.
4. Partially clock wise $\theta = \pi/4$.
5. Completely anti clock wise $\theta = -\pi/2$.

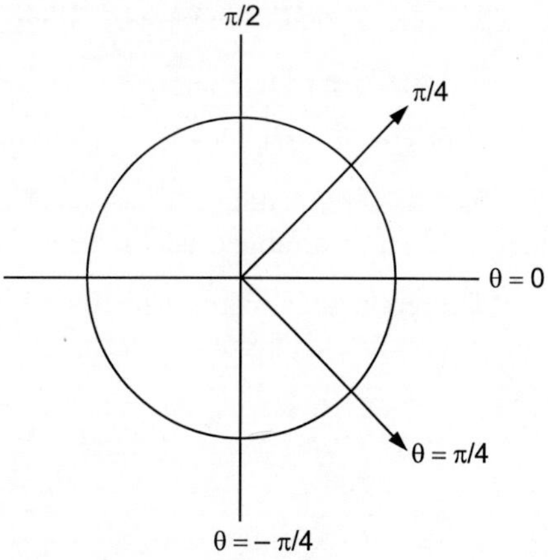

Fig. 7.5

In many optical experiments, colours are to be distinguished form one another. To form an angular fuzzy set a model defines the colours as per the following.

1. Absolutely Violet $\alpha = \pi/2$.
2. Very Violet $\alpha = 3\pi/8$.
3. Violet $\alpha = \pi/4$.
4. Fairly Violet $\alpha = \pi/8$.
5. in decided $\alpha = 0$
6. Fairly blue $\alpha = -\pi/8$.
7. Blue $\alpha = -\pi/4$.
8. Very blue $\alpha = -3\pi/8$.
9. Absolutely blue $\alpha = -\pi/2$.

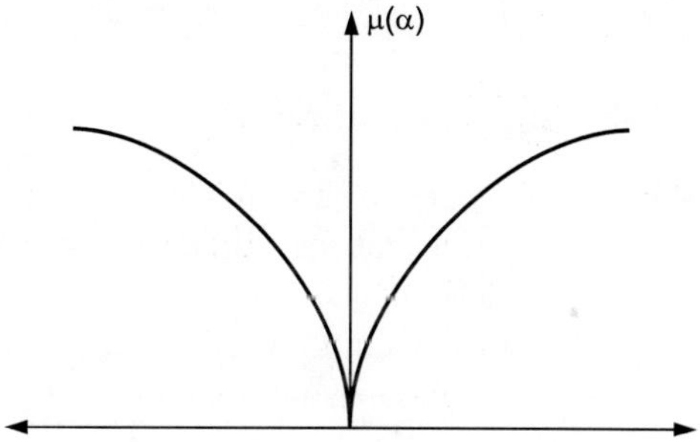

Fig. 7.6

7.2.4 Membership Function Through Neural n/ws

1. A neural *n/w* can be very conveniently used for determination of membership functions by suitable adjustments of weighting functions.
2. Neural system solves problems by adapting to nature of data signals they receive. One of the ways to achieve this is to use a training data set and a checking data set for i/p and o/p data signals.
3. Starting with a random assignment of weights to the element in different layers, the i/p from the training data set is passed through the neural *n/w* which computes a value of $f(x)$ output.
4. This is compared with the actual value $f(x)$ act
5. The error measure $\delta = f(x)_{\text{actual}} - f(x)_{\text{output}}$

   This is the error measure associated with the last layer of the neural *n/w*.

7.3 DEVELOPMENT OF RULE BASE AND DISCUSSION

Fuzzy Rule Base

Fuzzy linguistic descriptions are formal description of systems made by using IF – THEN rules. Knowledge about the system is encoded in the form.

IF (set of conditions) are satisfied THEN (set of consequents) can be inferred.

IF (x is A_1, x_2 is A_2... x_n is A_n) Then (y_1 is B_1, y_2 is B_2... y_n is B_n).

where linguistic variables x_i, y_i take value of Fuzzy sets A_i and B_i respectively.

Example: "If there are good results. There must be high placement". Here good and high are Fuzzy sets qualifying the variables such as results and placement.

The collection of rules in reference to a particular system is called the Fuzzy rule base. If the conclusion K is to be drawn from the rule base R (as conjunction of consequents K_i). we have

$$K = K_1 \cap K_2 \cap ...K_n \quad \text{so that}$$

$$u_k(y) = \min (\mu_{k1}(y), \mu_{k2}(y)... K_n(y)) \text{ A } y \in Y$$

Here Y is the universe of discourse.

If conclusion K is to be drawn from a rule base as a disjunction of individual consequents, we may note that

$$K = K_1 \cup K_2 \cup K_3 ... \cup K_n$$

$$U_K(y) = \max\{K_1(y), k_2(y)\}, A\ dy \subseteq Y$$

7.3.0 Defuzzification

The reverse process of fuzzification is defuzzification which converts a fuzzy set to a single crispvalue. It is like filtering out the vagueness and making out some meaning out of apparently contradictory ambigunous statements.

Of the many methods available for defuzzification, the most popular method are centriod method, centre of sums and mean of maxima: these can be summed up in the Table 7.1.

$$x^* = \frac{\sum_{i=1}^{N} x_i \sum_{k=1}^{M} u_{AK}(x_i)}{\sum_{i=1}^{N} \sum_{k=1}^{M} u_{AK}(x_i)}$$

Table 7.1

Centroid method	*COS*	*Mean of maxima*
$x^* = \frac{\int u(x)\, x dx}{u(x)\, dx}$ continous	$X^* = \frac{\Sigma_{i=1}^{N} xi\, \Sigma_{k=1}^{M} u_{Ak} - (xi)}{\Sigma_{i=1}^{N} \Sigma_{k=1}^{M} u_{Ak} - (xi)}$	$X^* = \frac{\Sigma x_i}{\lvert M \rvert}$
$x^* = \frac{\sum_{i=1}^{N} x_i\, u(x_i)}{\sum_{i=1}^{N} u(x_i)}$ descrete	M = No. of fuzzy sets	$M = \left\{ \frac{x_i}{u(x_i)} \text{ is height of set} \right\}$
N = No. of elements in the sample	N = Number of fuzzy variable	$\lvert M \rvert$ = cardinality of set M

Example:

	Area extent	*Shape*	*Area*	*x*
I	0–1	Δ	½ × 0.4 × 1 = 0.2	0.67
II	1–3.6	Δ	2.6 × 0.4 = 1.04	2.3
III	3.6–4	Δ	0.4 × 0.4 = 0.16	3.8
IV	3.6–4	Δ	½ × 0.4 × 0.2 = 0.04	3.8667
V	4–5.5	Δ	1.5 × 0.6 = 0.9	4.75
VI	5.5–6	Δ	0.5 × 0.6 = 0.25	5.75
VII	5.5–6	Δ	½ × 0.5 × 0.6 = 0.15	5.833
VIII	6–7	Δ	1 × 0.9 = 0.9	6.5
IX	7–8	Δ	½ × 0.9 × 1 = 0.45	7.33

$$x^* = \frac{\Sigma A\bar{x}}{\Sigma A} = \frac{\begin{matrix}(0.2 \times 0.67 + 1.04 \times 2.3 + 0.16 \times 3.8 + 0.04 \times 3.8667 + \\ 0.9 \times 4.75 + 0.25 \times 5.75 + 0.15 \times 5.833 + 0.9 \times 6.5 + 0.45 \times 7.337)\end{matrix}}{0.2 + 1.04 + 0.16 + 0.04 + 0.9 + 0.25 + 0.15 + 0.9 + 0.4}$$

$$= \frac{0.134 + 2.392 + 0.608 + 0.1546 + 4.275 + 1.473 + 0.8749 + 5.85 + 3.2985}{4.04}$$

$$= \frac{19.024}{4.04} = 4.70$$

7.3.1 COS

The overlapping area is counted twice.

The area of ~ A1 is ½ × 0.4 (3 + 5) from rule of trapezoids

The area of ~ A2 is ½ × 0.6 (4 + 2)

The area of ~ A3 is ½ × 0.9 × (3 + 1

So $$x^* = \frac{\frac{1}{2} \times 0.4(3+5) \times \left(\frac{5}{2}\right) + \frac{1}{2} \times 0.6(4+2) \times \left(\frac{3+7}{2}\right) + \frac{1}{2} \times 0.9 \times (3+1) \times \left(\frac{6+7}{2}\right)}{\frac{1}{2} \times 0.4 \times (3+5) + \frac{1}{2} \times 0.4 \times (4+2) + \frac{1}{2} \times 0.9 \times (3+1)}$$

$$= \frac{4 + 9 + 11.7}{1.6 + 1.8 + 1.8}$$

$$= \frac{24.7}{5.2} = 4.75$$

7.3.2 Mean of Maxima Method

As the aggregated fuzzy set is a continuous set, x^* the mean of maxima = 6 + 7/2 = 6.5 and height of the aggregated fuzzy set = 0.9.

7.3.3 Height Method for Neurofuzzy Application

This can be viewed as a two step procedure the consequent membership function C_i is converted into crisp consequent $Y = C_i$ where C_i is the centre of gravity. The centroid defuzzification is then applied to the rules with crisp consequents.

$$y = \frac{\sum_{i=1}^{m} w_i c_i}{\sum_{i=1}^{m} w_i} \qquad ...(7.3.1)$$

Where w_i is the degree to which the ith rule matches the input data. Neurofuzzy models use this type of defuzzification

This not only reduces the computation cost require by the previous two defuzzification techniques but also facilitates the application of neural networks into Fuzzy system. Many neuro Fuzzy models use this type of defuzzification method. The main disadvantages is that it is not well justified and is often treated as an approximate centroid defuzzification.

7.3.4 Other Methods

Centre of Largest Area

If the output fuzzy set has at least two convex sub-region the center of gravity (Z^*) is calculated using centroid method of the convex fuzzy sub-region with largest area is used to obtain the dufuzzified value Z^* of the output.

$$Z^* = \frac{\int z\mu_{c_m}(z)\, dz}{\int \mu_{c_m}(z)\, dz}$$

Where C_m is the convex sub-region having largest area. This condition applies in case the overall function is non-convex.

First (or) Last of Maxima

This method uses the overall output or union of all individual output fuzzy sets to determine the smallest value of the domain with maximized membership.

Steps:

1. $hgt\,(C_k) = \text{Sup}\ \mu_{Ck}\,(z)\ z \in Z \rightarrow C_K$ total set
2. $Z^* = \inf_{z \in Z} \left\{ z \in \frac{Z}{\mu_{CK}} (Z) = hgt\,(C_k)/ \right\}$

{sup = supremum, inf = infimum}

First the largest height in the union is determined. The first of maxima is given by the equation.

An alternative to this method is called Last of Maxima

$$Z^* = \inf_{z \in Z} \left\{ z \in \frac{Z}{\mu_{CK}} (Z) = hgt\,(C_k)/ \right\}$$

The choice of the methods depends on the following:

(*a*) Continuity
(*b*) Disambigity.
(*c*) Plausibility
(*d*) Computational simplicity

7.4 MEMBERSHIP EVALUTION OF GENETIC ALGORITHM

Genetic algorithm used to compute membership functions are provided with a functional mapping of system. Membership functions and these shapes are assumed for various fuzzy variables defined for a problem. These functions are then coded as bit strings and then concatenated. An evaluation function (fitness type) is used to evaluate the fitness of each set of membership function Parameters which can defined functional mapping.

7.4.1 Inductive Reasoning

Automatic generation of membership function can be made feasible by inductive reasoning which generic from specific this is performed by entropy minimization which "chusters" most optimally the parameters corresponding to output classes.

This techniques is based on an ideal scheme which describes the input/output relationships for a well established data base. This statement implies that in inductive reasoning, membership is generated based on solemly the data provided.

This method is quite useful for complete systems having abundant static data. In case of Dynamic data, this method fails as the membership function continually change with time.

Conclusion

It depends on the context of the problem.

1. The first criterion is continuity.
   The small change in input of fuzzy unit should not produce large change in output.
2. Disambiguity: defuzzification should result in unique value (ambiguous = vague).
3. Plausibility.
4. Computational Simplicity.
5. Weighting Method.

EXAMPLE 7.2:

1. $$Z^* = \frac{\int \mu(z).Z.dz}{\int \mu(z)\,dz}$$

Numerator is $\int_0^1 (0.3Z)\, ZdZ$

$$+ \int_1^{3.6} (0.3Z)\, ZdZ + \int_{3.6}^{4} \left(\frac{z-3}{2}\right) ZdZ$$

$$+ \int_4^{5.5} (0.5) ZdZ + \int_{5.5}^{6} (Z-5)\, ZdZ + \int_6^7 ZdZ + \int_7^8 (8-Z)\, dZ$$

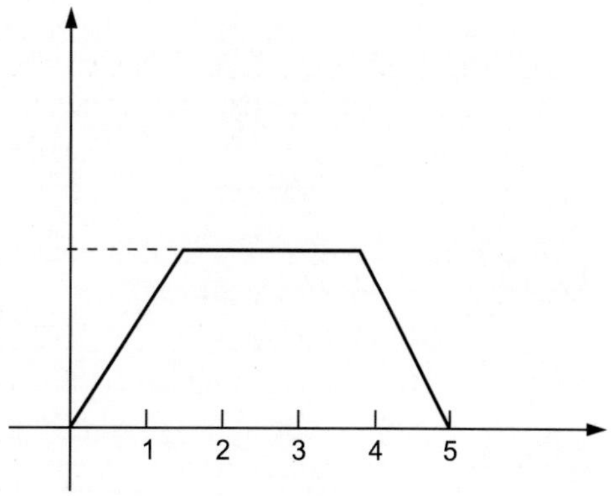

Fig. 7.7

$$\text{Denominator} = \int_0^1 (0.3Z)ZdZ + \int_1^{3.6} (0.3)dZ + \int_{3.6}^{4} \left(\frac{z-3}{2}\right) dZ + \int_4^{5.5} (0.5)dZ$$

$$+ \int_{5.5}^{6} (z-5)\,dZ + \int_6^7 dZ + \int_7^8 (8-z)\,dZ.$$

Solution: 1. Weight mean method (or) Average method

$$Z^* = \frac{\Sigma\mu_C(z)\cdot\bar{z}}{\Sigma\mu_C(z)} = \frac{0.3 + 2.5 + 0.5 \times 5 + 1 \times 6.5}{0.3 + 0.5 + 1} = 5.41\text{ m}$$

2. Mean maximum method:

$$Z^* = \frac{a+b}{2} = \frac{6+7}{2} = 6.5\text{ m}$$

3. Centre of sums method (COS)

$$Z^* = \frac{\int_z \sum_{k=1}^{n} \mu_C(z)\,dz}{\int_z \Sigma\mu_C(z)\,dz}$$

$$\frac{\int_0^8 [2.5 \times 0.5 \times 0.3(3+5) + 0.5 \times 0.5(2+4) + 6.5 \times 0.5 \times 1(3+1)]dZ}{\int_0^8 0.5 \times 0.3\,(3+5) + 0.5 \times 0.5(2+4)\,0.5 \times 1 \times (3+1)\,dZ} = 5.042$$

Example of Rule Based System

Operations like DIL; CON may be applied to large, major and small to express very large, more or less large, even if the continuous function appear to be linear (measured or assumed) they will not remain linear after such operation (the degree for overlap will change).

A Fuzzy generational statement may be multi exponent

IF A THEN B THEN C

Or $\quad R = A \times B \times C$

A combination of several conditional statements is required for optimum control of a real situation. When an actual input is available, the output is calculated from defuzzification. As it is impractical to have a explicit rule for every conceivable situation, a composition rule of inference may be adapted to produce a control rule. Assume on explicit control rule "if A then B". It may be inferred that if condition A' occurs it implies outcome B' where B' is inferred from a composition rule of inference of the type.

$$B' = A'_o R = A'_o (A \times B).$$

The designer decides that for a given input, the control algorithm checks if there exists a corresponding rule. If there is no such rule, then the rule in the immediate neighbourhood with a predetermined distance can be adopted.

7.4.2 Basic Design Steps

1. Identification of Inputs and outputs using linguistic variables
2. Assigning membership function to the variables
3. Building a rule base
4. Generating a crisp control action (defuzzification)

The linguistic variables membership functions and the rule base come from the experience. A large rule base leads to a smooth operation. Composition rule of inference can be used to trim the rule base. In Fuzzy system the rule base takes the form IF------ AND------OR -----, THEN -----, with operations AND OR being the Fuzzy Logic operation. In a non-Fuzzy system having a rule base control, one rule may operate at a time but with varied strengths. The mixture of rules operating with varied strengths leads to a crisp control action through the process of defuzzification. The process of defuzzification is not standard. Demanding on the need and suitability, any one of the methods discussed earlier may be adopted.

Let us take the case of a train approaching to or departing from a station. Inputs are distance from the station, speed of train. The output is the amount of broke power used. The membership functions for distance speed and brakes are shown.

7.5 RULE BASE

(*a*) If the speed is very slow and distance very—close, the brake power is very—light.
(*b*) If speed is slow and distance very—close, the brake power is heavy
(*c*) If speed is fast and distance very—close, brake power is heavy.
(*d*) If speed is very—fast and distance very—close, then brake power is very—heavy.
(*e*) If speed is very—slow and distance close, then the break power is light.
(*f*) If speed is slow and distance is close, then the brake power is light.
(*g*) If speed fast and distance is close, the brakes power heavy
(*h*) If speed is very—fast and distance is close, then brakes power.
(*i*) If speed is very-slow and distance far then is heavy brake power is light.
(*j*) If speed is slow and distance far, then brakes power is light.

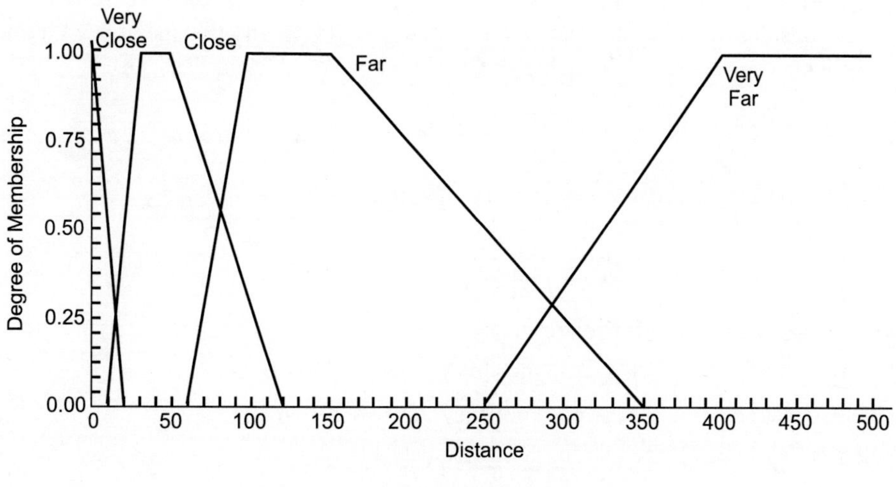

(*a*)

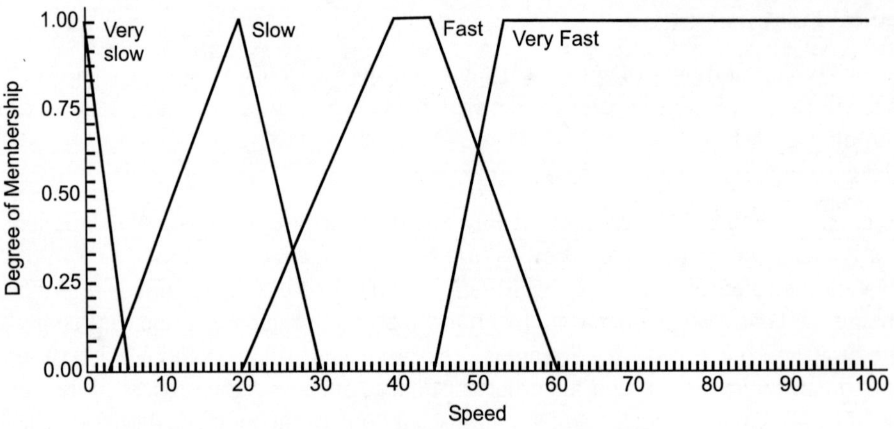

(*b*)

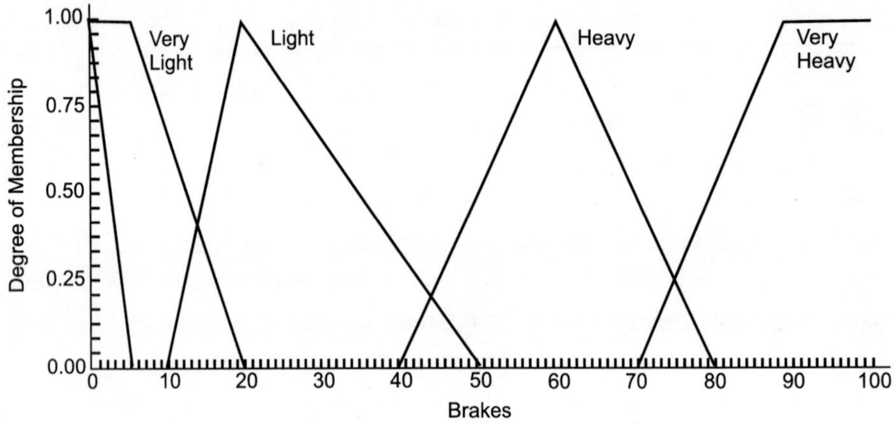

(*c*)

Fig. 7.8

These rules are summarized in Table.

Now, let us consider the control action if the distance is 100 m and the speed is 24.6 km/h. A speed of 24.6 km/h has a membership of 0.58 in Slow and 0.21 in Fast, i.e.,

$$U_{\text{slow}} = 0.58$$
$$U_{\text{Fast}} = 0.21$$

A distance of 100 m has a membership of 0.29 in Close and 0.88 in Far, i.e.,

$$U_{\text{close}} = 0.29$$
$$U_{\text{Far}} = 0.88$$

Accordingly, the rules that will fire are

- Rule #6 (slow/close) with a strength of 0.58 AND 0.29 = 0.29
- Rule #7 (fast/close) with a strength of 0.21 AND 0.29 = 0.21
- Rule #10 (slow/far) with a strength of 0.58 AND 0.88 = 0.58
- Rule #11 (fast/far) with a strength of 0.21 AND 0.88 = 0.21

Table 7.2 Input Matrix with Action. The Shaded Area gives the Action (Power of the Brakes)

Distance/Speed	*Very Slow*	*Slow*	*Fast*	*Very Fast*
Very Close	Light	Heavy	Very Heavy	Very Heavy
Close	Light	Light	Heavy	Very Heavy
Far	Light	Very Light	Light	Heavy
Very Far	Very Light	Very Light	Light	Light

If using min - max defuzzification, an OR operation follows, leading to

0.29 OR 0.21 OR 0.58 OR 0.21 = 0.58

The crisp control action will be to apply 58% of the maximum braking power. If using center-of-gravity defuzzification, the crisp output is determined by the weighted sum of the centroids of the membership functions. The centroid is the center of the output membership function adjusted to the degree of rate filing. The triangular membership functions become trapezoidal, as shown in Fig. 7.8.

The crisp control action is determined by the weighted average defined by:

$$\text{Crip action} = \frac{\sum^{n} (\text{rule strength})_n \cdot (\text{action})}{\sum^{n} (\text{rule strength})_n}$$

This leads to

$$\text{Brakes Power} = \frac{(0.29)(5) + (0.21)(70) + (0.58)(5) + (0.21)(30)}{0.29 + 0.21 + 0.58 + 0.21}$$

$$= 19.65\% \text{ of maximum braking power.}$$

7.5.1 Linguistic Hedges

Fundamental atomic terms can be altered by adjectives (nouns) or adverbs such as very, slight, fairly. Almost mostly these modifies are called linguistic hedges as the singular meaning of an atomic term is "hedged" from the original interpretation.

The fuzzy sets can be used as calculus of interpretation so that linguistic hedges can modify the membership for a basic atomic term. Let a basic linguistic atom α be subjected to some hedges and let it be defined as

$$\alpha = \int_{R}^{R} \alpha_x \frac{(x)}{r} \text{ then}$$

$$\alpha^2 = \text{“very } \alpha = \int \frac{[u_x(r)]^2}{r}$$

$$\alpha^4 = \text{very, very}$$

$$\text{“Pulse” } \alpha = \alpha^{1.25} \text{ “minus” } \alpha = \alpha^{0.75}$$

$$\text{“Slightly” } \alpha = \sqrt{\alpha} = \int_{R}^{R} \frac{[u_x(r)]^{0.5}}{r}$$

These can be again correlated with concentration/dilation or intensification as discussed earlier.

7.5.2 Fuzzy Sets and Natural Language

Fuzzy sets can be used as a calculus for interpretation of natural language. Despite vagueness and ambiguity, natural language is the means of human communication. It is proper that a mathematical theory which deals with fuzziness and ambiguity is used to express the linguistic base of languages.

Use of natural language in the expression of a knowledge forms gives rise to rule based systems. The decomposition of compound rules into canonical rule from as logical propositions can be extrapolated.

The expression of rules as a collection of logical implications can be manipulated by some inferencing schemes which can be graphical in nature also.

According to cognitive scientists humans base their thinking on patterns and images rather than on numerical. Natural language refers to mental images with vague but simple terms. Inspite of intrinsic vagueness, human communicating in a common language have very little difficulty in fundamental understanding. Zadeh conceptualized language as a shell of own thoughts. (The use of fuzzy sets as the mathematical foundation of natural language is the obvious conclusion of the above analysis).

The used language can be broken into some fundamental terms and certain linguistic connectors for these terms. Fuzzy sets can be used in the numerical description of the terms and in the prescription of connection of terms into strings of intelligible forms of expression.

The fundamental terms can be called atoms and collection of atoms can be considered to form molecules or phrases examples of atomic terms are slow young handsome a collecting atomic terms is called a composite examples being very slow vehicle , slightly young person's handsome actor sets of atomic terms can exists as elements and the universe of natural language terms can be thought of as universe A another universe B can represent the universe of cognitive interpretations' (or meanings)

It is rather straight forward to imagine a universe of terms however, it is difficult to develop α universe of interpretation such a universe can be a collection of cognitive patterns and mental images and the corresponding interpretations can be represented by fuzzy sets. So, an atomic term or linguistic variable can be interpreted using fuzzy sets.

If we define a specific atomic term in the universe of natural language, A as element α and we define a fuzzy set *S* in the universe of interpretation or meanings *R*, as α specific meaning for the terms α natural language can be a sort of mapping. This mapping M is from a set of atomic terms in *A* to a corresponding set of interpretation defined on *R*. Each atomic term α in *A* correspond to a fuzzy set *S* in *R* which is the interpretation of α. Thus, this mapping can be denoted by $M(\alpha, s)$.

The fuzzy set *S* representing the fuzziness in mapping between an atomic terms and its interpretation can be represented by a membership function $u_u(\propto, r) = u_s(r)$

7.5.3 Predicate Logic

In this logic, constants stand for objects which do not vary or change (Ex: Rod. Pen, Phone) variables are symbols which have values of the objects are qualified by the qualifier associated with the variables (Ex: *a*, *b*, *c*, *x*, y, *z*). Predicates represent association between objects which are variables and have truth of values normally, a predicate has a name representing the relation followed by its arrangements standby the objects of association (ex: writes (Hari. *X*) implies that hari weites anything). If hari wrote on any topic writes (Hari, *x*) acquires a true value for *x* being qualified quantifies can ∀ (All) and some (∃)*V* is universal quantifier where as ∃ existential qualifier.

Ex: man (*x*): *X* is man

Mortal (*x*): *X* is mortal

All men are mortal can be represented as :

$$\forall x\ (\text{man}(x) \rightarrow \text{mortal}\ (x)$$

An existential quantifier has conjunction where as universal qualifier is with implication logical formula can be qualified by multiple qualifiers

Functions are similar to predicates inform and in the representation of association. The difference between predicates and function is that the former acquires thruth values only where as the functions can have values other then truth values. Functions basically describe the object.

Ex: Brother (Ram) (Ram's Brother)

A Simple proposition is a statement which is either "True" or "false" but not both. It is also called an atom. A sequence of propositon can be linked by using connectives or operations, There are five major operators.

Symbol	*Connective*	*Remarks*
$\wedge$	AND	BINARY
$\vee$	OR	BINARY
$\neg$ or –	NOT	UNARY
$\Rightarrow$	IMPLIES	BINARY
=	EQUALS TO	BINARY

In case of $\Rightarrow$ the proposition proceeding is called antecedent and that following is called consequent. The semantics or meaning of logical connectives can be explained through truth table having rows as interpretations.

P	Q	$P \wedge Q$	$P \vee Q$	$\sim P$	$P \Rightarrow Q$	$P = Q$
T	T	T	T	F	T	T
T	F	F	T	F	F	F
F	F	F	F	T	T	T
F	T	F	T	T	T	F

Any logical formula comprising n propositions will be associated with 2^n interpretation. If a formula has all true interpretation it known as tatology and a formula which all false interpretation is called contradiction?

Example: Evaluate if $(A \Rightarrow B)$ $(B \Rightarrow A) = (A = B)$ can be a tautology

A	B	$A \Rightarrow B$	$B \Rightarrow A$	X $(A \Rightarrow B) \wedge (B \Rightarrow A)$	$Y A = B$	$X = Y$
T	F	F	F	F	F	T
F	T	F	F	F	F	T
T	T	T	T	T	T	T
F	F	T	T	T	T	T

The formula can be considered to be tautology

Example: IF P THEN Q ELSE R should be expressed by

IF P THEN Q can be expressed as $A \Rightarrow B$

If P THEN R can be expressed as $A \Rightarrow C$

The equivalent expression is $(A \Rightarrow B) \vee (A \Rightarrow C)$

In classidical logic based on rules of propositional type the logic connectives can be explained as:

Connectives	*Symbols*	*Example*
AND	$\wedge$	$P \wedge Q$ P and Q are both true
OR	$\vee$	$P \vee Q$ either is true
NOT	$\neg$	$\neg P$, P not true
IMPLICATION	$\Rightarrow$	$P \Rightarrow Q, \neg P \vee Q$ P implies Q is true
EQUALITY	$=$	P is equal to Q
EXCLUSIVE OR	$\oplus$	$P \theta Q = (\neg P \wedge Q)\ (P \wedge \neg Q)$

Properties of Propositional Logic

Commutativity
$$P \vee Q = Q \vee P$$
$$P \vee Q = Q \wedge P$$

Associativity
$$P \vee (Q \vee R) = (P \vee Q) \vee R$$
$$P \wedge (Q \wedge R) = (P \wedge Q) \wedge R$$

Distibuitivity
$$(P \vee R) \wedge R = (P \wedge R) \vee (Q \vee R)$$
$$(P \vee Q) \wedge R = (P \wedge R) \wedge (Q \vee R)$$

Idempotence
$$P \wedge P = P$$
$$P \wedge P = P$$

Identitiy
$$P \vee \text{True} = \text{True}$$
$$P \vee \text{False} = P$$
$$P \wedge \text{True} = P$$
$$P \wedge \text{False} = \text{false}$$

Absorption law
$$P \wedge (P \wedge Q) = P$$
$$P \vee (P \vee Q) = P$$

Involution
$$\neg(\neg \text{P}) = P$$

DeMorgan's Law
$$\neg(P \wedge Q) = \neg P \vee \neg Q$$
$$\neg(P \vee Q) = P \vee \neg Q$$

7.5.4 Inference in Propositionallogic

Inference mechanism provides the advantage to a sum conclusion from a given set of premises or facts S_1..........S_n. From the statements "if it is match day there will be no traffic on the road and "today is Monday" "inference is that the traffic will be quite high on the road". The three important rules are:

- Modus Ponens
- Moduls tollens
- Chain rule

Modus Poneins

Given that $P \Rightarrow Q$ and P is true, then Q is true

$$\begin{array}{c} P \Rightarrow Q \\ \underline{P} \\ Q \end{array}$$

Modus Tollens

Given that $P \Rightarrow q$ and $Q \Rightarrow R$ is true then $P \Rightarrow R$ is tue

$$\frac{\begin{array}{c}P \Rightarrow Q \\ \neg Q\end{array}}{\neg P}$$

Chain Rule

Given that $P \Rightarrow Q$ and $Q \Rightarrow R$ are true $P \Rightarrow R$ is true

$$\frac{\begin{array}{c}P \Rightarrow Q \\ Q \Rightarrow R\end{array}}{P \Rightarrow R}$$

Chain rule is a reflection of transitivity for the connective $\Rightarrow$.

7.5.5 Predicate Logic

Prepositional logic may not lead to a correct conclusion always. If we take the proposition S_1 and S_2 and conclusion e as per the following:

S_1: Girls play good Badminton

S_2: Saina is a girl

S: Saina is good at badminton

In this case the conclusion seems to be right. But in place of Saina, if it is sania mirza, the inference could be wrong.

To make predicate logic effective we have to divide it into subjects and predicates. In a propositon x is R when x is a subject, R is predicate which represents the property, or quality. We can generalise by taking any subject x over specified universe of discourse and predicates can be more correctly substituted by a function $R(x)$ defined over universe of discourse. Predicates can also represent the association between subjects or objects. "Jolly loves poems" can have Jolly as subject, poems as object and loves as a predicate. This can be represented by loves (Jolly, poems).

Binary predicates can be extended to by quantifies such as universal quantifiers and existential quantifiers $\exists$. Jolly loves some poems can be $\exists\, x$ poems $(x) \Rightarrow$ loves (Jolly, x) Ruby likes all books can be $\forall\, x$ book $(x) \Rightarrow$ likes (Ruby, x).

Fuzzy Propositional Logic

Fuzzy logic can have truth value in between 0 and 1 "Jolly some what likes novels" can have a degree attached of the order of 0.5. Such a statement is a fuzzy proposition. The truth value for fuzzy proposition can be "TP" fuzzy logic basically generalises the idea of truth values which makes it effective tool for getting inferences from the facts which are partially true. Such feature of fuzzy logic is called approximate reasoning:

Fuzzy Connectives

Symbols	*Connectives*	*Examples*
$\wedge$	Conjunction	$P \wedge Q \equiv \text{Min}\ (T_P, T_Q)$
$\vee$	Disjunction	$P \vee Q \equiv \text{Max}\ (T_P. T_Q)$
$\neg$	Negation	$\neg P \equiv 1 - T_P$
$\Rightarrow$	Implication	$P \Rightarrow Q \equiv \text{Max}\ (1 - T_P T_Q)$

Fuzzy Quantifiers

1. **Absolute:** Quantifies defined on S that characterise linguistic terms such as at least 100, about 100, around 100, much more than 100, very much more that 100. etc..
2. **Relative:** Quantifies defined on [0, 1] and characterise linguistic terms such as about half, almost all, very nearly half.

7.6 TYPES OF FUZZY RULES

Two types of fuzzy rules are prevalent:

1. Mapping rules
2. Implication rules

The Mapping rule describes functional mapping between inputs and outputs using linguistic terms. Implication rule discribes a generalised logic implication relationship between two logic formulae based on linguistic and fuzzy linguistic terms.

7.6.1 Fuzzy Mapping Rules

Crisp values are converted to fuzzy so that these can be used for decision-making on human level. In real life cases, it is required to find out a relationship between a set of observable values to a set of values which can be specified only Vaguely. In control systems observable state variables are mapped to the state of approximate control actions with the mapping rules. Most industrial applications are based on fuzzy mapping.

7.6.2 Fuzzy Implication Rules

These are generalization of Boolean logic to multivalued logic. Using Modus ponens and modus tollens.

Example: IF Dr. Rao is going abroad and he has got visa or he has got departmental permission. Then he may be going to the U.S.

Desirable fuzzy inferences can be derived from fuzzy implication by some other criteria.

(*a*) Generalised *moduls* ponens
(*b*) Generalised *moduls* Tollens
(*c*) Mismatch
(*d*) Fundamental chaining

Generalised Modus Ponens

Lingustic hedges are used to modify the antecedent of a fuzzy rule

(*a*) X is $A \Rightarrow y$ is B

$$\frac{x \text{ is very } A}{y \text{ is very } B}$$

(*b*) X is $A \Rightarrow y$ is B

$$\frac{x \text{ is almost } A}{y \text{ is almost } B}$$

Mismatch

In two valued logic, if the antecedent is known to be false, inference cannot be made if the consequent is true or false

Generalized Modus:

(a) X is $A \Rightarrow y$ is B

$$\frac{x \text{ is not (very } B)}{x \text{ is not (very } A)}$$

(*b*) X is $A \Rightarrow y$ is B

$$\frac{x \text{ is not (almost } B)}{x \text{ is not (almost } A)}$$

Generlised Modus Ponens

(*a*) X is $A \Rightarrow y$ is B

$$\frac{x \text{ is very } A}{x \text{ is very } B}$$

(*b*) X is $A \Rightarrow y$ is B

$$\frac{x \text{ is more or less } A}{x \text{ is more or less } B}$$

Chaining: This is same as the chaining rule is classical logic

$$x \text{ is } A \Rightarrow y \text{ is } B$$

$$\frac{x \text{ is } B \Rightarrow Z \text{ is } C}{x \text{ is } A \Rightarrow Z \text{ is } C}$$

Fuzzy Implication Functions

(*a*) Arithmetic

$T(P \Rightarrow Q) = (1 - (T_P + T_Q))$

(*b*) Maximum

$T(P \Rightarrow Q) = \text{Max } (1 - T_P \min (T_P, T_Q)$

(*c*) Standard Sequence:

$$t(P \Rightarrow Q) = \begin{matrix} 1 & t_p < t_q \\ 0 & otherwise \end{matrix}$$

(*d*) Godelian sequence

$$T(P \Rightarrow Q) = \begin{matrix} 1 & t_p \leq t_q \\ t_q & otherwise \end{matrix}$$

(*e*) Goguen's formulation

$$t(P \Rightarrow Q) = \begin{matrix} 1 & t_p \leq t_q \\ \frac{t_q}{t_p} & otherwise \end{matrix}$$

Fuzzy Inference System

Fuzzy inference systems are of two types:

(*a*) Non additive rule based systems

(*b*) Additive rule based systems
Mamdami's fuzzy inference system is of the first type Takagi Sugeno-Knag (TSK) system and Koso's are of the second type

Mamdani's Fuzzy Inference System

It *A* as proposed by mamdani in 1975 as an attempt to control a steam-engine-boiler combination by synthesing a set of linguistic control rules obtained from experienced human operators.

In this system. The main steps are:

1. Fuzzification of input variables using the input membership function
2. Rule evaluation (finding the consequence of rule by combining the rule strength and the output membership function).
3. Aggregation of the rule outputs (combining the consequence to get an output).
4. Defuzzfying the output.

SUMMARY

- This unit discusses Fuzzy logic components including predicate logic and defuzzification.
- Examples of defuzzification has been dealt with in simple fashion.
- He IF-THEN rules also have been considered for development of Fuzzy logic system.

Short Answer Questions

1. Define Fuzzification?
2. What is meant by Membership function?
3. What are the features of membership function?
4. Define De-Fuzzification?
5. What is meant by Inductive Reasoning?
6. List the basic design Steps?
7. Explain Predicate logic?
8. What are fuzzy implication functions?

Long Answer Questions

1. Explain about fuzzy rule based system with example
   Refer 7.4
2. Define defuzzification? Explain briefly different method of Defuzzification?
   Refer 7.3
3. What is meant by MOM Method explain it briefly
   Refer 7.2

4. What is meant by membership? Explain about types of membership functions?
   Refer 7.2
5. What are basic design Steps. Explain them with an example?
   Refer 7.5
6. Write a short note on
   (*i*) α-cuts of Fuzzy set
   (*ii*) α-cuts of Fuzzy Relation
7. Explain about the types of Fuzzy rules?
   Refer 7.6

❑❑❑

8

APPLICATIONS OF NEURAL NETWORKS AND FUZZY LOGICS

8.1 INTRODUCTION

As indicated earlier Neural Networks have wide ranging applications some of which are

(*a*) Process identification and control
(*b*) Fault diagnosis
(*c*) Load or forecasting
(*d*) Bioinformatics
(*e*) Knowledge extraction
(*f*) Health care
(*g*) Image analysis (patterns recognition) and interpretation
(*h*) Signal analysis and interpretation
(*i*) Intrusion detection
(*j*) Communication
(*k*) Efficient design of RF/wireless circuits
(*l*) Image processing and compression.

For Example: Human face detection in visual sense is also being carried out.

For the purpose of this course the first three will be touched upon. This rather endless list is only a sample of areas where ANNs have been applied. In days to come the growth will be vertical as well as horizontal.

8.2 PROCESS

This is used to obtain expected plant (process to be regulated) ouput signal. Such identification is very much essential for adaptive system. Basically there are two types of identification

(*a*) Forward plant identification
(*b*) Plant inverse identification

In forward plant identification, training involves identification of error signal as difference of desired response and output of the network. The weights are adjusted to minimize the error. Network is designed by mapping plant input to plant output such that measurements of input and output are carried out simultaneously. The past inputs of the plant are obtained through delay elements.

In the plant inverse identification, the output is fed to the neural network as the input the weight adjustment is done as per the difference of the input and output.

The limitation of plant inverse identification exists when the plant is not mapping defined and input output relation has multiple option for non-existence of plant inverse characterizes, mapping may drift to the average of the multiple possibilities.

8.3 DYNAMICS OF PLANTS

In this case the sequential plant outputs are used along with the most recent output to create the current state (Fig. 8.1) The output vectors of the plant and time delays have to be considered for complete evaluation. Time between sampling of inputs used for training and of control of plant is measured.

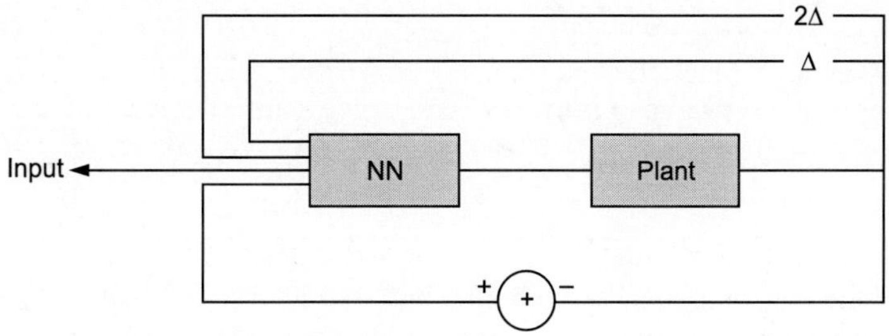

Fig. 8.1

8.3.1 Limitation of On-Line Learning

If the plant is known completely, online learning system based on back propagation does not work properly. In the cases in which error is only known to the plant, the back propagation method cannot be properly used.

Let S be the input to neural network which produces range of outputs $O_1........O_n$ which are fed to the plant giving outputs as $y_1..........y_m$. The error vectors are $S_1....y.........S_m ...y_m$.

According to back propagation rules

$$\Delta W_{ij} = -n \frac{\delta E}{\delta w_{ij}}$$

$$dW_{ij} = -n \frac{\delta E}{\delta o_i} \frac{\delta o_i}{\delta w_{ij}}$$

$$i = 1...n, j = 1, m$$

Error is defined as $E = \frac{1}{2} \sum_{i=1}^{m} (S_i - y_i)^2$

The error plant output is transmitted back to plant input and $\frac{\delta E}{\delta o_i}$ is computed from:

$$-\frac{\delta E}{\delta o_1} = (s_1 - y_1)\frac{\delta_1}{\delta o_1} + (s_2 - y_2)\frac{\delta y_2}{\delta o_1} + ... + (s_n - y_n)\frac{\delta y_m}{\delta o_1}$$

$$-\frac{\delta E}{\delta o_2} = (s_1 - y_1)\frac{\delta y_1}{\delta o_2} + (s_2 - y_2)\frac{\delta y_2}{\delta o_2} + ... + (s_n - y_n)\frac{\delta y_m}{\delta o_2}$$

$$-\frac{\delta E}{\delta o_m} = (s_1 - y_1)\frac{\delta y_1}{\delta o_n} + (s_2 - y_2)\frac{\delta y_2}{\delta o_n} + ... + (s_n - y_n)\frac{\delta y_m}{\delta o_n}$$

This matrix gives $m \times n$ Jacobian matrix. When the plant is unknown perturbation method may be adopted for evaluation . Approximate values can be obtained leading to a major limitation of the method.

8.3.2 Inverse Identification

Here S is the plant input and y is the neural network input as contracted to forward plant identification. During training, the plant output generates response only as desired response is not presented.

Back propagation techniques are used for minimization of error. The network is designed in a way to implement the mapping of plant of inverse. The network is trained by implementing the property of plant inverse.

In actual practice, forward plant implementation is only practicable.

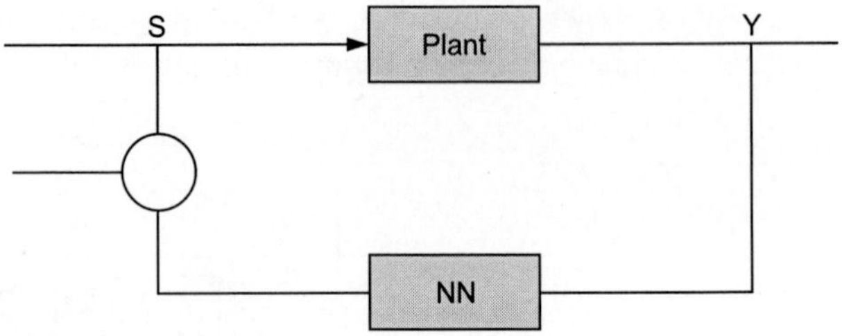

Fig. 8.2

Static Plant

Indirect learning architecture or model based architecture is used in static plants with slight modification. The method of two copies of neural network is changed to the method of only one copy. The main aim is to obtain the response of the plant when the input is not known.

8.4 IDENTIFICATION PROCESS

From control systems, it is known that plant is a physical process to be regulated. The expected plant output signal is obtained by plant identification. In controls of adaptive systems, identification of plant is very much essential for evaluating plant variations with time.

The neural networks used for plant identification are multilayer feed forward architecture with Error Back Propagation principle.

8.4.1 Forward Plant Identification

The neural network receives the same input as the plant. The response of plant output should match the response of the neural network.

During the process of identification the weights are adjusted to minimize error vector through Error Back Propagation the network is designed by mapping plant input to plant output with simultaneous measurement.

Plant identification can be achieved by many complex process models the past inputs can also be considered using delay elements.

8.5 INVERSE IDENTIFICATION

In a conventional control system, the output of the process under control is measured and converted to a signal representing the output $c(t)$. Comparing it to a reference signal $r(t)$ and generating an error single $e(t)$, control action is computed and applied.

In proportional control $p(t) = K_p\, e(t)$ where K_p is referred to as a gain. In integral control, $I(t) = K_1 \int_o^t e(t)dt$ where K_I is constant. In derivative action, $D(t) = K_D\, d\frac{e(t)}{dt}$, K_D being constant. When all action are combined, the overall control signal $m(t)$ can be written as

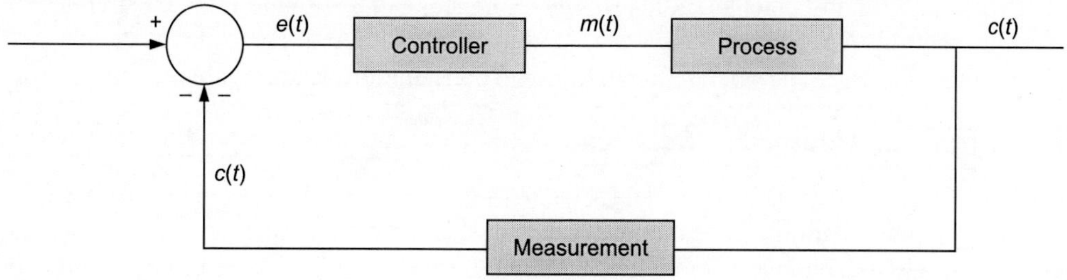

Fig. 8.3

$$m(t) = K_p e(t) + KI \int_o^t e(t)dt + K_D d\frac{e(t)}{dt}$$

The concept of PID controller can be extended to digital control systems where a difference, rather than differential equation describes the operation:

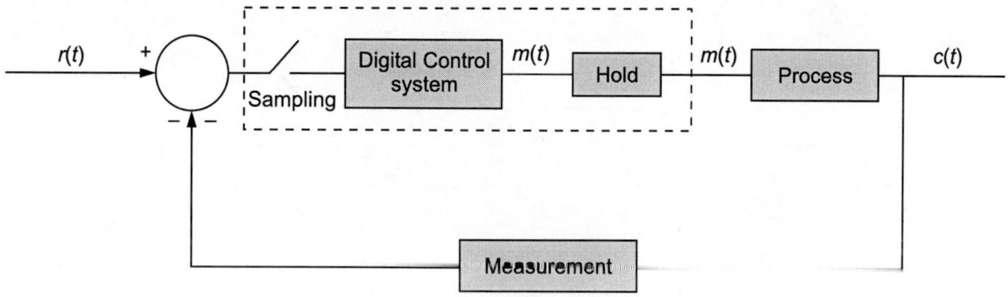

Fig. 8.4

8.6 NEURAL NETWORK FOR INDUCTION MOTOR CONTROL

An induction motor has highly non-linear torque- speed characteristics. The aim is to maintain constant torque at variable speed known as vector control of induction motor. Using the concept of field oriented control, three phase currents are vectorised into two orthogonal axes (direct and quadrature). Variation of current in these axes control torque and flux separately.

Neural network is applied to estimate torque, flux and flux angle of the motor at any instant of time comparing the torque and speed with desired set values error signal is generated to compensate the difference.

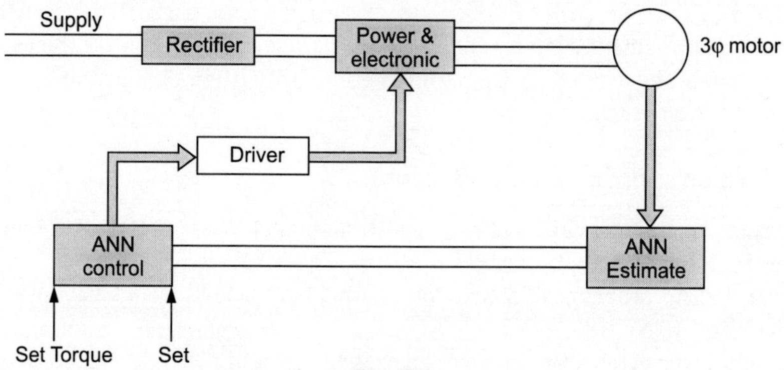

Fig. 8.5

A DSP estimator or any other suitable (proven reference) can be used to train a neural network for the desired estimation. A reference model can be used to train a neural controller to decide the firing angle to minimize the error and operate the motor with constant torque at variable speed.

Block Diagram of Neural FD Scheme

The BACK PROPAGATION is used for fault diagnosis in case of 3-tank system consists of 9 neurons in the input layer and 2 neurons in the output layer. Single and double hidden layers are the options for this case. The single hidden layer can be tried with 5, 10 or 15 layers. In case of double hidden layer there can be (5, 10), (15, 25) (15, 5) and (20, 5) neurons.

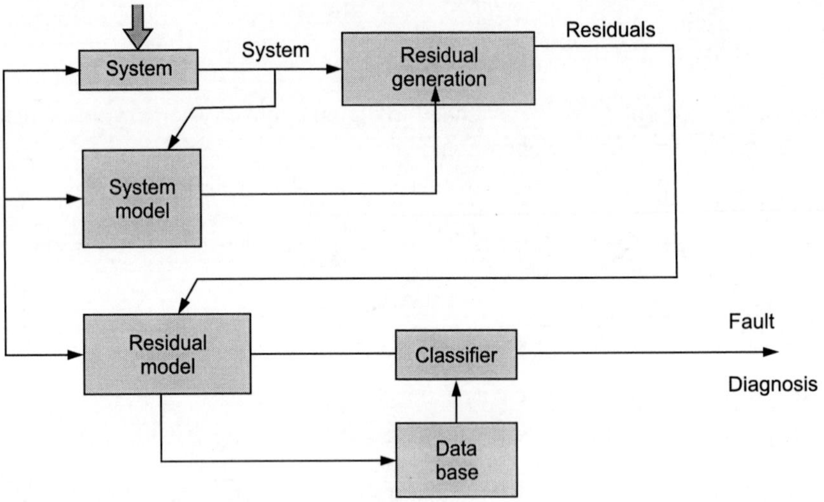

Fig. 8.6

Some squared error is found to be the least and the fault classification performance is more accurate for the structure with double hidden layers consisting of 20 and 15 neurons.

(a) *Input Processing Field:*

It is divided into two portions:

1. Input portion.
2. Interface portion.

The input portion presents the input vector given. The interface portion combines the signal form *i/p* with that of F_2 layer. This is used in comparing the similarity of input signal to the *wt* vector for the cluster, which is selected for learning.

8.6.1 Load Forecasting in Power System

1. Multilayer perceptions, Hopfield N/W and Kohonon N/W are preferred for analysis of power system. Multilayer perception are also preferred as MLP.
2. In power systems problems related to forecasting, fault detection and diagnosis is are important application of ANN.
3. Using multilayer perceptron, the emphasis is of sigmoid f^n.

4. A traditional perceptron consists of two layers they are input and output.
5. This structure as limitation delaying with non-linear.
6. Introduction of hidden layers with input and output makes more versatile. According to Kolmogarov theorem, the three layer perception allows as to approximate any non-linear F^n.
7. MLP is a universal approximation.
8. The sigmoid F^n is a smooth F^n which can be differentiated weights in order to.
9. Back propagation algorithm is conveniently used with sigmoid f^n for determination of optional *wt* of A MLP.
10. It is to develop Hopfield N/W.
11. It converges over optional solution.
12. In the area of power system, the problems may be of difference weights. The problem formulation may be expressed in terms of

    (*a*) non-linear (*b*) large scale
    (*c*) dynamical (*d*) discrete
    (*e*) stochastic random life (*f*) fuzzy period and various parameter system

    Among these factors, the non-linear and larger scale system needs more parameters to solve.
13. ANN is used to solve complex power system model through MLP, Hl & KN.

Description of Load Demand Pattern

The factors effecting load system demand are:

1. Trend effects
2. Cyclic time effects
3. Weather effects
4. Random effects like human

So, the load profile is economic in nature and temporal seasonal and annual.

Load forecasting is one of the most successful applications of ANN in power system. Neuro fuzzy (*NF*) computing is a popular framework for solving complicate problems of power systems. If we have knowledge expressed lignuistic rules we can build in Fuzzy interface system and if we have data or learn simulation from training. For building an affairs Fuzzy set, Fuzzy operators and knowledge.

Formulation may be Expressed in Terms of

(*a*) Non-linear (*b*) Large scale
(*c*) Dynamical (*d*) Discrete
(*e*) Stochastic random life (*f*) Fuzzy period and various parameter systems.

The architecture of MLP has been used for forecasting. There are two hidden layers, within the 1st hidden layer the processing is done by a non-linear function, usually by a hyperbolic tangent. And 2nd, the processing is done by the 2nd hidden layer using sigmoid f^n. The training sample data has to be fairly large to contain all the required information.

$$\text{The mean absolute \% error} = \frac{1}{N}\sum_{i=1}^{N}\frac{P_{acti_i} - P_{redict_i}}{P_{acti_i}}$$

Back propagation algorithms are quite slow for practical problems. High performance algorithms, which can cover 10 to 100 times faster than BPN, are used. The main categories of fast algorithm are

1. Heuristic technique (variable learning rate BPN or resilient BPN).
2. Numerical optimization techniques such as conjugate gradient and Quasi Newton.

MLP structure in a particular case:

The features are:

Neural network	*MLP*
Number of hidden layers	1
Number of hidden neurons	17
Activation f^n in hidden layers	tan-sigmoid
Activation f^n used in output layer	pure linear
Mean absolute percentage error (MAPE)	0.18

$$y_k = f(\mu_k)$$

MLP in Forecasting

The idea behind the use of MLP models in load forecasting is simple: it is assumed that future load is dependent on past load and external factors (i.e. temperature), and the MLP network used to approximate this dependency. The inputs to the network consists of those temperature values and past load values and the output is the target load values(for example a load value of a certain hour, load values of many future hours, the peak load of a day, the total load of a day etc). Therefore, the building of a MLP model for load forecasting can be seen as a non linear statement identification problem. The determining of the model structure consists of selecting the input variables and deciding the network structure. The parameter estimation is carried out by training the network on load data of the history. This requires choices concerning the learning algorithm and appropriate training data. The model validation is carried out by testing on load data, which has not been used in training.

However, the modeling with neural network is different to modeling with linear system models. The nonlinearity and great adaptability of the network models make it possible to use specific indicators as input variables. In the case of load forecasting, the hour of the day and day type of the target hour, for instance, can be included as binary codes in the network input. The network model can be understood to be based on pattern recognition functions, where different input patterns are mapped in different ways this makes the models very different to, for example, ARIMA models , which assumes that the load time series can be made stationary(i.e., invariant with respect to time) with suitable filters. The handling of the special load condition is easier for neural network models then for ARIMA models.

Another matter supporting neural network models is relatively rapid changing of the characteristic in the load behavior. This is a problem with statistical models, because they cannot always keep up with the sudden change in the dependence of the load. For example, the beginning of holiday season etc., can change the load behavior rapidly. As neural network models are in essence based on pattern recognition functions, they can in principle be hoped to recognize the change condition without re-estimating the parameters. This requires of course that conditions corresponding to the new situation has been used in training, and that network inputs contains the information necessary for recognizing the condition.

On the other hand, a problem with MLP models is the black box like description of the dependencies of the future values on the past behavior. The understanding of the model is very difficult; the common sense can hardly be applied in order to see how the outputs depend on inputs. This responding of the model to an input pattern, which is very different to any experienced during the learning, can be unexpected. This can happen in new conditions, even if the model is validated with test data.

EPRI Load Forecasting Engine

The Electric Power Research Institute (EPRI) developed a load-forecasting engine using 24 neural networks. The networks are grouped into four categories depending on the time of the day. The categories are:

Category 1: Nine neural networks; Forecasts the load between 1 AM and 9 AM.

Category 2: Nine neural networks; Forecasts the load between 10AM and 2PM, and between 7PM and 10PM.

Category 3: Four neural networks; Forecasts the load between 11PM and 12PM.

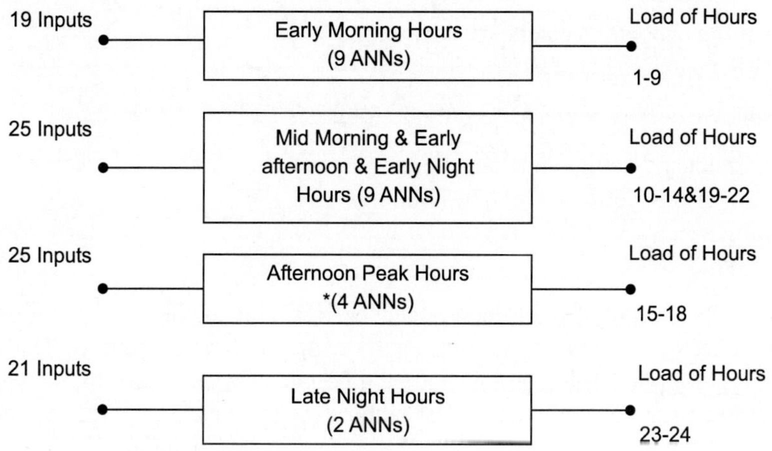

The categories have different inputs that are listed below:

Category 1: Forecast for early morning general input.

- Load in the last four hours.
- Temperature in the last three-four hours.

Category 2: Forecast for off peak hours.

- General input.
- Temperature of previous hours.
- Previous day's load and temperature of hours close to this hour.

Category 3: Forecast for afternoon peak hours.

- Temperature of previous hours and forecasted temperature of future hours close to this hour.
- Yesterday's load and temperature close to this hour.

Category 4: Forecast for late night hours.

- General input.
- Forecast temperature next four hours.

The general inputs are again in five subdivisions as given below:

- Same hour-load, temperature and humidity, from the previous day. (3)
- Same hour-load, temperature and humidity of the second previous day. (3)
- Same hour-load, temperature of the past week. (7 × 2)
- Same hour forecasted temperature and relative humidity of the next day. (2)
- Day of the week index: Sunday: 01, Monday: 02, etc., (1)

The numbers within the brackets indicate the number of input in each sub-category. The load forecasting engine has one output for the hourly load, for 24 hours. For extended forecasts we make use of the pre-forecasted values. That is, for two day ahead forecast, the values obtained for one day ahead forecast is used. Also there is a constant updating in the forecasted value, as the latest load and weather data become available. As a result of this constant updating process, the neural network is trained continuously and the weights are adjusted daily. This retraining uses the most recent load and weather data, thereby tracking the changes in weather and load pattern. As discussed earlier, the load during the holidays has different pattern and is slightly reduced. Due to the insufficiency in the count of historical data for this period, the forecasts are usually inaccurate. One way of going around the problem was to consider the holiday as:

1. Saturday, if the shopping centers are open or,
2. Sunday, if the shopping centers are closed.

In essential the weather service has to provide forecasts for:

- Daily maximum and minimum temperature.
- Daily maximum and minimum relative humidity.
- Rain and fog.
- Maximum wind speed.

But no hourly data are provided for the weather services. Due to this dependence on weather, EPRI developed another ANN engine for forecasting hourly temperature and humidity. This second network has 28 inputs and 24 outputs. The inputs include the hourly temperature of the previous day and the extreme temperatures of the past two days. The output was the expected hourly temperatures.

The accuracy was 3% for the next day forecast and 7%-8% for longer terms. The performance was concluded to be satisfactory. The input data and various plots of interest are depicted alongside. The input data is given in table3. The utility is the day (Sunday: 01, Monday: 02, etc.,), the period is the days; in the year 1996, for which the forecast was made and the days ahead indicate the duration of the forecast.

SAMPLES OF PERFORMANCE OF SECOND GENERATION ENGINE WITH ONLINE WEATHER FORECAST INPUTS								
		Days Ahead						
Utility	*Period in %*	*1*	*2*	*3*	*4*	*5*	*6*	*7*
1	8/21-11/19	3.22	4.09	4.32	4.68	5.18	5.77	6.41
2	10/26-11/25	2.54	3.14	3.53	3.94	4.43	5.16	5.72
3	10/26-11/25	2.47	3.28	3.66	4.07	4.57	4.87	5.64
4	9/5-11/19	2.15	2.78	2.95	3.26	3.79	4.38	4.89
5	2/1-11/20	2.81	3.99	4.65	5.25	5.98	6.53	7.07

The forecast of the network is compared with the actual load in the plot below. The dark lines represent hourly forecasted-load is while the actual hourly load is depicted by the dotted lines.

The plot proves that the accuracy of the network is indeed good. The plot of forecasted hourly peak load vs. the actual values is shown below:

Load Forecasting

MLP is used as one of tools deals with time series analysis of a load. As applications of ANNs to load forecasting, short-term load forecasting is of concern. High accuracy of the load forecasting improves security and generation cost. However, the forecasting problem is not so easy because of the complicated factors such as nonlinearity and random-like behavior of system loads and weather conditions. The conventional methods, say, ARIMA, Kalman Filtering, and multiple regression models lack the adaptivity of model structure in case where unexpected factors such heat or cold waves influence significantly system loads. It is expected that MLP allows us to carry out adaptive forecasting appropriately.

The KN feature mapping permits us to classify load patterns and serves as their pre-filtering technique. After classifying load patterns with KN and constructing ANNs for load forecasting in each cluster, we can have more exact prediction. This technique is very useful at a turning point of season where there is significant uncertainty of weather.

Security Assessment

MLP is used to provide an approximate solution. Static and dynamic security assessments often require on-line computation. However, the conventional analytical methods are inclined to be time-consuming. In order to evaluate the solution efficiently, the nonlinear mapping of MLP is utilized to reduce the computational burden and deal with the characteristics of power systems. That allows us to carry out on-line monitoring/assessment in transient, small signal stability, and voltage instability.

KN applications may be divided into a couple of approaches. One is that the output neuron represents power system conditions to handle the security assessment. In other words, the neuron implies a qualitative measure, say, secure or insecure. Some output neurons correspond to secure power system conditions while the others do to insecure ones. The other is that each output neuron has some information on a qualitative index for small signal disturbance analysis or voltage instability. Namely, information on power system conditions is embedded in each output neuron on the feature mapping. Once an output neuron is fired, information on the index is recalled from the database.

Typical problem characteristics of load forecasting, security assessment and fault detection/diagnosis:

Features	*Load Forecasting*	*Security Assessment*	*Fault Detection Diagnosis*
No. of Input Variables	Several	Many	Many
No. of Out Variables	One	One/Many	Many
Alternative Analytical Approach	Available	Available	Not Available
Necessity of On-Line Evaluation	X	XX	X
Typical Problem to be Addressed	Fast Learning	Topology changes	Topology Changes
Size of ANN	Small	Large	Large
Direct Implementation of ANN*1	Acceptable	Not Easy	Not Easy

Note: symbol X denotes that the feature is necessary while Symbol XX means that the feature is highly necessary.

*1: Direct implementation means that ANN is applied to the problem without sophisticated feature extraction of input variables.

8.7 FUZZY CONTROL

8.7.1 Introduction

A process can be controlled by an operation sufficiently without the need of any mathematical model for the system. The best example is that of a driver controlling a car without the mathematical model of driving. A driver knows how to reach the destination safely. "How to drive" is based on the ability to interpret linguistic statements about the system to reason qualitatively. Supposing we are too close to the vehicle ahead, we slow down our own vehicle a bit. "Slow a bit", "too close" are Fuzzy terms. The final (resultant) action is not Fuzzy but very specific like pressure on the brake. The value of pressure selected by the driver is based on his criteria of driving. The situation becomes complicated if there are many conditions to consider before the action. For instance, if we are getting close, we slow down a bit and if the weather is foggy, we tend to slow down even more. The resultant action (pressure on the brakes) comes as a conclusion of selecting a single value from the Fuzzy set which comes from consideration of the two conditions. The basic steps are—one to infer the possible actions and other to select one particular, suitable action. Often, during driving we respond with clear definite action to a crisp (clear and definite) situation. When we see a signal, we stop. Red signal and STOP can be described as Fuzzy singletons. Thus we may conclude that, without being aware of position, we have been driving by Fuzzy techniques all these years.

Fuzzy controllers are known to perform better than the conventional PID or DDC controllers particularly when noise becomes an important factor.

8.7.2 Review of Conventional Systems

In a typical automatic control system, the output of the process render control is measured and converted into an electrical signal which represents the output $c(t)$. A comparison is made between $c(t)$ and input signal $n(t)$ as reference the error signal generated actuates the controller to generate a control action signal $m(t)$.

Such systems are designed on mathematical models based on the process and specifications. These models can be produced and implemented with quite good precision.

For a manually controlled system, similar actions take place rather qualitatively. As an example of the car again, a driver "calculates" the distance the car ahead with eyes acting as tranduler and compares this distance with that set in his brain. The difference signal activates the control action of slowing down using legs(considered to be "manual"). If the experience of the driver is wider, the driving will be proportionately smoother. Intelligent drivers can infer actions based on the driving experience to handle situation which did not occur earlier.

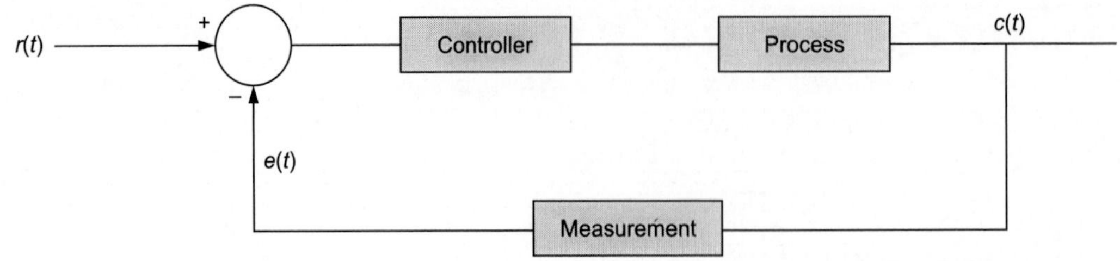

Fig. 8.7

Basically, there are two important aspects to be specified are determined for a given system:

1. **Transient Response:** This is related to the system stability. A system is said to be stable when its output does not increase without limit with time (and there is no oscillation with time).

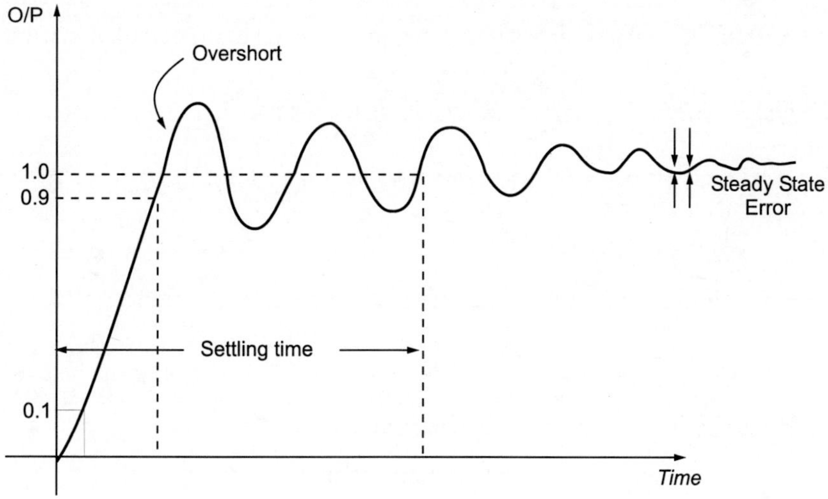

Fig. 8.8

The figure shows you an example of transient response and the parameters such as rise time, over shoot and settling time.

2. **Study State Response:** This is related to accuracy of the system and can be assessed by measuring the steady state error. In designing, digital systems offer an easy solution as most of the digital circuits such as microprocessors and memories are predesigned process. In many practical situations having constraints, digital systems fail to implement all the relevant factors.

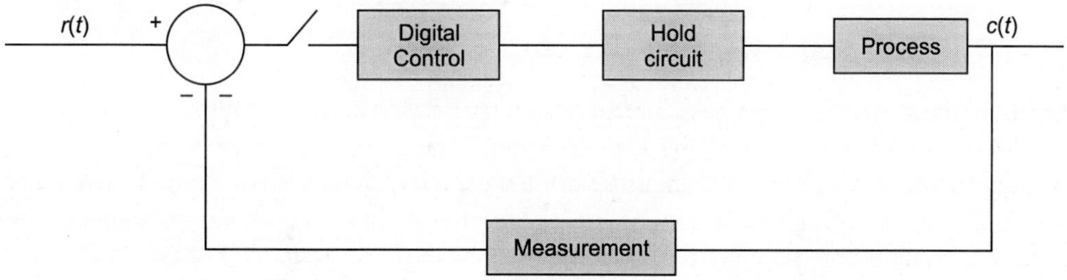

Fig. 8.9

The analog input is sampled at regular intervals (sample periods). The sampled error signal $e^*(t)$ and control signal $m^*(t)$ are important controlling elements. The sampling frequency has to be at least two times the highest frequency of the original signal otherwise the samples cannot represent correctly. Delay of computation, sets the upper limit for the sampling requency.

8.7.3 PID Controller

The most popular industrial control system is the proportional-and derivative controller known as PID controller. The control signal has three different configurations. The basic objective is to bring down the overshoot, setting time, steady error to the values specified and maintain the stability of the system. The configurations are summarized below:

(*a*) **Proportional Action:** The control signal $P(t)$ is proportional to the error signal at the present moment in time: $P(t) = K_p\, e(t)$ where K_p is called gain.

(*b*) **Integral Action:** Here the control signal $I(t)$ is proportional to the cumulative values of the error upto the present $I(t)^1 = K_1 \int_0^t e(t)\, dt$ where K_1 is a constant.

(*c*) **Derivative Action:** Here the control signal $D(t)$ is proportional the derivate of the error signal $D(t) = \frac{de(t)}{dt}$ where k_D is a constant.

With combination of (*a*), (*b*) and (*c*), the overall control signal $M(t)$ can described by $M(t) = K_p e(t) + K_i \int_0^t e(t)\, dt + K_d \frac{de(t)}{dt}$. The concept of PID controller can be extended to DDC by using difference equations and delay elements.

$P(T_k) = K_p\, e(t_k)$ where t_k is the sampling moment and integer $I(t_k) = K_I\, h$ $[e(0) + e(h) + e(2h) + \ldots\ldots\ldots\ldots]$ $K = 0, 1, 2, \ldots\ldots\ldots\ldots$ The derivative action is $D(t_k) = K_D \left[\frac{e(t_k) - e(t_k - 1)}{h}\right]$ the overall $M(t)$ is the combination of the above digital expression.

Only PI control are suitable for process where the dynamics are essentially of first order (PID is for 2nd order). For example, level control intakes can be handled by PI where as motion with friction can be handled by PID.

8.7.4 FLC (Fuzzy Logic Control) System

FLC describes the algorithm for process control as a Fuzzy relation between information on the condition of the process to be controlled and the control action. The basic difference from the conventional control algorithm is that information (linguistic or Fuzzy model) about the system replace a mathematic model. It is also distinct from the expert system as the rules of FLC system are not obtained from the human expert through the system but formulated clearly by the FLC designer. Fuzzy expert systems are also available.

The similarity between FLC and conventional system is that a control action is produced based on measurement of output in conventional system, control action is reached through an algorithm based on multiplication by a constant, derivative integration or a combination of all three. The basic premise of Fuzzy control algorithm is a conditional statement between the Fuzzy input variable *A* and a Fuzzy output variable *B*. This is expressed as a linguistic implication statement: *A-B* (condition *A* implies condition *B*).

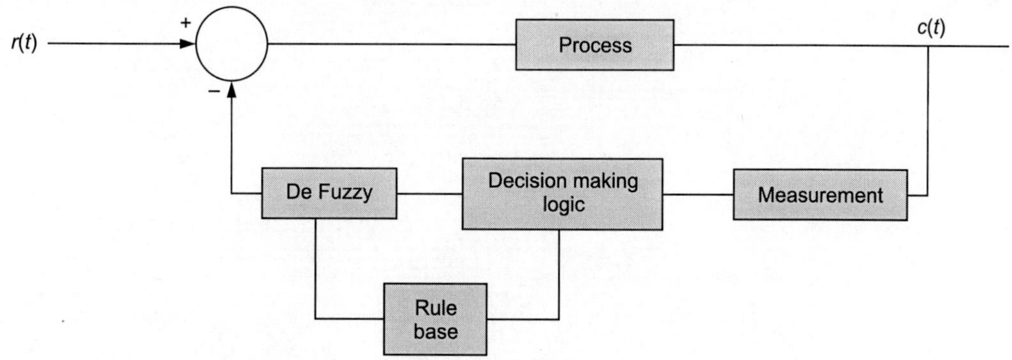

Fig. 8.9

This can be written as: IF A THEN B. This expression is similar to that Cartesian multiplication ($R = A \times B \equiv$ If A Then B. A Fuzzy variable is expressed through a Fuzzy set which is defined by the membership function μ. The Fuzzy variable may be continuous or discrete continuous. Variable can be quantized and expressed *as if it* was discrete.

Fuzzy Logic Control vs PID Control

One may still have the desire to describe the PD, PL or PID algorithms linguistically to enhance the performance of an existing system. One can use rules for PD such as

- IF error is positive AND change in error is almost zero, THEN the control action is positive.
- IF error is negative AND change in error is almost zero. THEN the control action is positive.

For PID you can use rules of the type

- IF error is almost zero AND the change in error is positive AND the sum of errors is positive, then the control action is positive.

The AL220 Fuzzy Controller (see Appendix B) was shown to be able to create an approximation of a PID controller [15]. The ease of modification it allows can be used to improve performance in specific applications. For example, it makes the incorporation of extra sensors in the system easy: just add inputs from sensors, define the membership function for these inputs, and add Fuzzy variables to the Fuzzy rules. Since Fuzzy Logic can accommodate non-linear sensors, the AL220 can be programmed to accommodate low-cost non-linear sensors. Adding the fact that the AL220 chip is inexpensive, PID controllers can be implemented at low cost using Fuzzy Logic techniques.

8.7.5 Industrial Applications of FLC

One of the early applications of Fuzzy Logic control occurred in 1978 in Denmark. Where a Fuzzy Logic controller was used on a cement kiln. The system was then marketed by F.L. Smith & Co. of Copenhagen in 1980. Since then numerous Fuzzy Logic control systems have been reported in a wide range of applications that include industrial processes, transportation systems, robotics, and consumer products. In the following sections we give a bird's eye view of some classical examples of industrial applications of Fuzzy Logic Control (FLC).

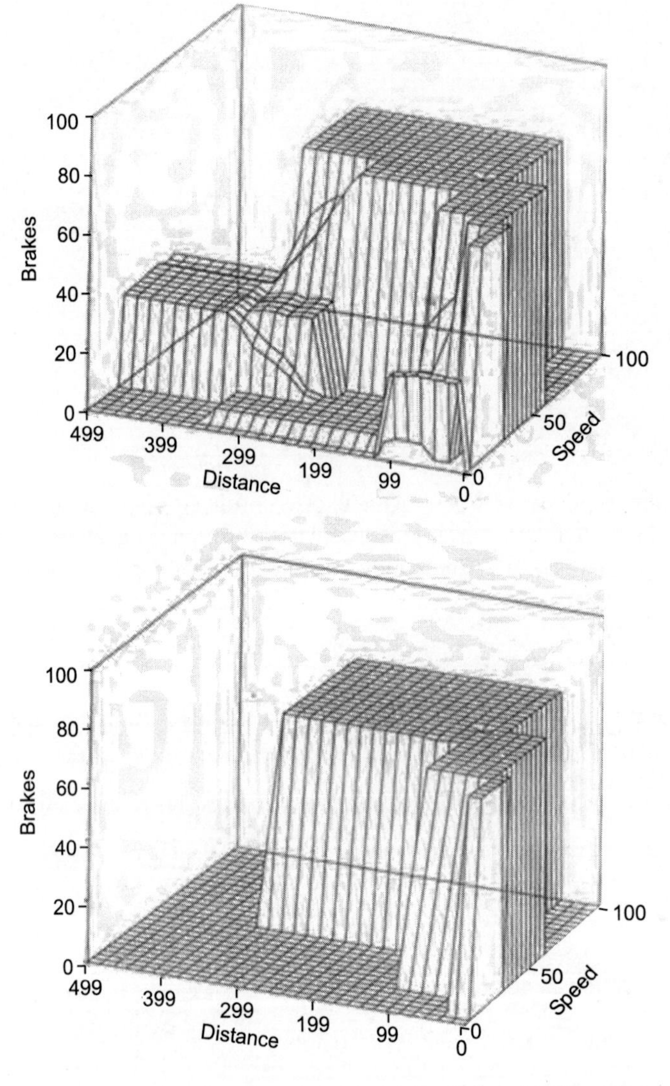

Fig. 8.10 *Results of Simuation using (a) Six Rules, (b) Three Rules.*

A. Cement Kiln

A cement kiln is a rotating chamber in which ground limestone and clay react with each other at temperatures of 1000-1400°C. The objective, of the control operator is to maximize the output product, minimize fuel consumption, and keep the machinery running within specified limits.

The Fuzzy Logic controller reported by Holmbald and Ostergaard was realized with a minicomputer. The Fuzzy control algorithm was based on the operators' experience and field tests. Examples of Fuzzy Logic control rules are:

- If *TC* is *ZE* and *FL* is *LO*, then *FA* is *MN*
- If *TC* is *ZE* and *FL* is *OK*. then *FA* is *ZE*
- If *TC* is *SN* and *FL* is *LO*: then *FA* is *SP*

Where *TC* = kiln drive torque change, *FL* = free lime, *FA* = fuel flow adjustment, *ZE* = zero, *LO* = low. *OK* = okay, *SN* = small negative, and *SP* = small position

The Fuzzy variables *LO*, *HI*, etc. were defined in the algorithm. The extent combination of each control rule is judged based on the actual conditions of the kiln. The FLC was reported to reduce fuel consumption slightly compared to the operator controlled system.

B. Heat Exchanger Process

The core of a heat exchanger is two hydraulic systems: hot and cold water curriculum water in a reservoir is heated electrically. It is then pumped into a pipeline to the neighbourhood of a cold water pipe. The cold water gets heated and the hot water loses heat. The hot water then goes back to the reservoir for reheating.

The control problem reported was to adjust the hot water flow rate FH, and the reservoir heating power, W, so that the cold water outlet temperature, TCOS and hot water inlet temperature, HI, take on setpoint values TCOS and THIS respectively. Two 7 × 9 decision tables were needed. Using an IBM/ 1800 computer 7 hours of computing time were used.

As an example of control rules, one of the rules for controlling the cold water outlet TCO states that:

- If TCOE is *LP* or *MP* and THIE is NOT *LP* or *MR* then *FK* is *MP* where

  LP = large positive, *MP* = medium positive.

The experimental results showed characteristics comparable to those of PI controller.

C. Boiler and Steam Engine

The steam engine speed was controlled by adjusting the throttle opening at the input of the engine cylinder, while the boiler pressure was controlled by the heat input to the boiler.

The operating experience and technical knowledge of the process was user to specify the control rules. The algorithm was implemented on a PDP8 computer

As an example of control rules,

- If *PE* = *N* and then if CPE = NOT (*NB* or *NM*) then *HC* = *PB*
- If *PE* = (*NB* or NM) then if CPE = *NS* then *HC* = *PM*

Where *PE* = pressure error, CPE = change in pressure error, *HC* = heat input change, *NB* = negative big, *NM* = negative medium, *PB* = positive big, *NS* = negative small, *PM* = positive medium.

When FLC was compared with DDC, the FLC system was found to be mean less sensitive to process parameter changes.

D. Water Purification Plant

In a water purification plant, raw water is purified by injecting chemicals at rate related to water quality. The feed rate of chemicals is judged by a skilled operation Aluminum Sulphate or PAC (Polymerized Aluminium Chloride) is used as a coagulant Aluminium Sulphate is less expensive than PAC but is not as effective in low temperature water. An alkaline agent, such as line or caustic soda, is used for pH adjustment. A chlorine agent is used for sterilization. All chemical agents, except the coagulant agent are usually fed at a fixed rate that is proportional to the following raw water. The reaction of the coagulant with water impurities is a complicated process to model; no universally accepted method of a feed rate exists.

Yagishiia et al. structured a system based on Fuzzy Logic so that the feed rate of the coagulant can be judged automatically even by an unskilled operator Ten control rules were used. As an example,

- If TUUP = *MM*, then DDOS = *PS*
- If TUUP = *ML*, then DDOS = *PM*

Where TUUP = grade of water, DDOS = amount of compensation dose, *MM* = medium, *PS* = positive small, *ML* = medium large, and *PM* = positive median.

The Fuzzy variables *MM*, *PS*, etc. were assumed to have a normal distribution. A particular set was defined by specifying the average value and the standard deviation.

The system was used successfully in a water purification plant in Akita Civ in Japan in 1983.

Refuse Incineration Plant

In a refuse incineration plant, a refuse supply feeder takes the refuse to a stoker in which refuse is burnt. The major purposes of the combustion control are:

1. Providing adequate supply of refuse (oversupply extinguishes the flame and insufficient supply leads to poor combustion).
2. Keeping the evaporation rate and furnace temperature at their specified values. The Fuzzy control for the plant in Shizuoka City reported by Owner consists of Fuzzy sensors, Fuzzy controllers, and conventional combustion control (ACC). A total of 45 control rules based on the knowledge and experience of the operator were incorporated. For example.

- If the evaporation rate is in the target zone, then leave it to automatic operation by ACC.
- If the evaporation rate is low and the refuse layer thickness is high, the increase amount of combustion air and decrease feeder speed.

E. Arc welding robot

A hazardous environment surrounds the arc-welding process. Very bright welding sparks, high temperature, and toxic fumes are associated with the process. It also takes a long time to train a human welder. Thus, the process is a suitable field of application for a robot.

The system has two inputs and two outputs representing the composition of a vertical and a horizontal control system. The FLC was designed for the two system turns independently. Eight control rules were used. The membership function of the Fuzzy variables PH and NH were defined by empirical formulae.

The experimental data indicated that using FLC, the system produced less vibration of tracking locus compared to the case where PI controller was used.

Stability

Traditionally, stability analysis is a fundamental aspect in determining the reliability of a control system. Conventional stability analysis is based on the availability ability of a mathematical model of the system to be controlled. Difficulty arises in Fuzzy control systems where the major advantage is in devising a control system without the need for a mathematical model.

Mamdani voiced the opinion that a Fuzzy controller can be analyzed qualitatively to assure that a runaway instability does not occur. Changes to the control system structure to improve performance can be introduced by monitoring the system running open-loop.

For example, King and Mamdani reported on temperature control of stirred tanks. Their initial results showed a tendency to oscillate around the set point. Simulation of the process was carried out to determine the cause of the instability. Time delay was observed to be a primary cause of the instability. An effective way to solve the problem was to modify the universe of discourse by changing the quantization levels, and by adjusting the overall loop gain. The results were confirmed to be satisfactory on the real process.

Mamdani suggested that the discussion on stability is irrelevant if it implies that no attempt be made to control difficult processes unless a rigorous theory can be found to design Fuzzy Logic controllers. This was accepted implicitly by numerous reports in the literature of Fuzzy Logic systems design. The concern about a systematic method of studying the stability, however, existed since the early stages of Fuzzy Logic control applications. Nevertheless, the literature has much work reporting on applications but much less reporting on stability studies. (see for example references).

If we have a mathematical model, we expect a mathematical analysis of stability. But if we have a linguistic model, why should we expect mathematical analysis of stability? Would it be possible to have a linguistic stability criterion? Stability can be defined linguistically of course. Probably the closest analysis to such an idea is the one by Kiszka et al.

The basic equation of a simple Fuzzy dynamic system has the form

$$X_k + 1 := X_k \circ U_k . R \qquad k = 0, 1, 2, 3, \ldots$$

where X_k and $X_k + 1$ are the Fuzzy sets of the states at k^{th} and $(k-1)^{\text{th}}$ time instants, respectively, U_k is the input at the kth instant, and R is the Fuzzy relation between the input and output.

A free or unforced dynamic system is realized by setting the input to zero. In this case, we have $X_k + .J = X_k \circ P$

where $P = U_k \circ R$ for U_k = zero for all k, $k = 0, 1, 2, \ldots$

A state of X_5 of a free dynamic system is an equilibrium state if $X_k + 1 - = X_k$ for all k, i.e.,

$$X_s = X_S \circ P$$

Various authors put forward algorithms to calculate the greatest equilibrium state. Kiszka et al. suggested an algorithm based on the energy of the system. A dynamic system is stable if its total energy is minimal and constant. The system is unstable if its energy increases with time, and oscillatory if its energy fluctuates periodically. They proposed an intuitive measure of energy of a Fuzzy relation. It relies on the most salient physical characteristics of the Fuzzy relation, such as the position in the support set, maxima of membership function, etc. They defined the energy of a Fuzzy relation P as

$$E(P) = \frac{1}{n.m} \sum_{i=1}^{n} \sum_{j=1}^{m} W(x_1 y_1).f(x_1 y_1)$$

where $w(x_1, y_1)$ is the function responsible for the position of the Fuzzy relation in the universe of discourse, $F(U_p(X_i, Y_j)$ is the function responsible for the shape, spread, etc., of the Fuzzy relation, and m and n are the cardinalities.

Further, they defined the incremental rate of energy change or the relational characteristic energy function, as

$$\Delta W(p, i) = E(P^i) - E(P^{i-1})$$

where $F = P \circ P \circ P \ldots P$ is the z'-times max-min composition of the Fuzzy relation P.

8.8 FUZZY CLASSIFICATION

An example of fuzzy classification is the fuzzy *K*-nearest neighbourhood algorithm. Given a set of classified date the *K*-NN algorithm determines the classification of an input based on the class labels of the K-closest neighbour in the classification data set.

The chief requirement of such techniques is in supervised pattern recognition in which data with known classification are used to determine the classification of new data.

In unsupervised pattern recognition, training data are not necessary. For image processing the availability of training data (relevant knowledge) has major implications. Firstly, it implies that one or a set of processed images (for training) is not difficult to be obtained. Secondly, it implies that the general content of the images such as objects and regions can be anticipated before processing. For instance, the general content of MRI image is predictable even though there are many variations in shape and volume.

8.8.1 K-NN Method

The main advantage of K-NN is the computational simplicity. Its problems are of two types:- firstly, each of the neighbours is considered equally important in determining the classification of input data. Secondly, the algorithm only assigns a class to the input data. Because of these limitations, fuzzy K-NN algorithm was developed.

The fuzzy K-NN algorithm assigns to an input x a membership vector ($\mu c_1(x)$, μc_2 (x)... $\mu c_p(x)$), where p is the total number of classes. The class membership are calculated based on the formula:

$$\mu_c i\ (x) = \frac{\sum_{j=1}^{k} \mu c_i(x_j) \dfrac{1}{\left\|x - x_j\right\| 2/m - 1}}{\sum_{j=1}^{k} \dfrac{1}{\left\|x - x_j\right\| 2/m - 1}}$$

X_1, X_2............X_k denote the K-nearest neighbour of X.

The parameter m serves as a parameter such that as $M \rightarrow \infty$, the term $\|x - x_j\|$ $2/m - 1$ approaches one regardless of the distance. So the neighbour X_j are more evenly weighted. As m decreases the closer neighbours are weighted for more hearily than those which are away. The factor effectively reduces the number of neighbours which contribute to the membership value of the input data.

SUMMARY

- This unit discusses some of the numerous applications of NN and FL .
- From motor to load forecasting number of practical cases have been discussed.
- Comparison with existing conventional system has been carried out.
- Scope of further work is also been indicated

Short Answer Questions

1. How many types of identification are there and what are they?
2. What is meant by Input processing field?

3. What are the factors affecting the Load system demand?
4. Explain about Transient Response?

Long Answer Questions

1. What is meant by dynamic plant? Explain about limitation of online learning?
   Hint: Refer 8.3
2. Explain about inverse identification process?
   Hint: Refer 8.4
3. Explain about PID controller and differentiate fuzzy ligic control and PID control?
   Hint: Refer 8.7.3
4. Explain about the neural network for induction motor control?
   Hint: Refer 8.6
5. What is meant by
   (*a*) Transient Response?
   (*b*) Study state Response?
   Hint: Refer 8.7.2
6. Explain the Industrial application of FLC?
   Hint: Refer 8.7.5
7. What is meant by K-NN meathod?
   Hint: Refer 8.8.1

❑❑❑

APPENDICES

APPENDIX 1: HODGKIN-HUXLEY MODEL

The most successful and widely-used models of neurons have been based on the *Markov kinetic* model developed from Hodgkin and Huxley's 1952 work based on data from the *squid giant axon*. We note as before our voltage-current relationship, this time generalized to include multiple voltage-dependent currents:

$$C_m \frac{dV(t)}{dt} = -\sum_i I_i(t, V)$$

Each current is given by *Ohm's Law* as

$$I(t, V) = g(t, V) \cdot (V - V_{eq})$$

where $g(t, V)$ is the *conductance*, or inverse resistance, which can be expanded in terms of its constant average $\bar{g}$ and the activation and inactivation fractions m and h, respectively, that determine how many ions can flow through available membrane channels. This expansion is given by

$$g(t, V) = \bar{g} \cdot m(t, V)^p \cdot h(t, V)^q$$

and our fractions follow the first-order kinetics

$$\frac{dm(t, V)}{dt} = \frac{m_\infty(V) - m(t, V)}{\tau_m(V)} = \alpha_m(V) \cdot (1 - m) - \beta_m(V) \cdot m$$

with similar dynamics for h, where we can use either τ and m_∞ or α and β to define our gate fractions.

With such a form, all that remains is to individually investigate each current one wants to include. Typically, these include inward Ca^{2+} and Na^+ input currants and several varieties of K^+ outward currents, including a "leak" current. The end result can be at the small end 20 parameters which one must estimate or measure for an accurate model, and for complex systems of neurons not easily tractable by computer. Careful simplifications of the Hodgkin-Huxley model are therefore needed.

Hodgkin-Huxley Experiments: In 1952, Hodgkin and Huxley wrote a series of five papers that described the experiments they conducted that were aimed at determining the laws that govern the movement of ions in a nerve cell during an action potential. The first paper examined the function of the neuron membrane under normal conditions and outlined the basic experimental method pervasive in each of their subsequent studies. The second paper examined the effects of changes in sodium concentration on the action potential as well as the resolution of the ionic current into sodium and potassium currents. The third paper examined the effect of sudden potential changes on the action potential (including the effect of sudden potential changes on the ionic conductance). The fourth paper outlined how the inactivation process reduces sodium permeability. The final paper put together all of the information from the previous papers and turned them into a mathematical models.

Hodgkin

Huxley

A. L Hodgkin and A.F. Huxley developed a mathematical model to explain the behavior of nerve cells in a giant squid in 1952. Their model, which was developed well before the advent of electron microscopes or computer simulations, was able to give scientists a basic understanding of how nerve cells work without having a detailed understanding of how the membrane of a nerve cell looked. To create their mathematical model, Hodgkin and Huxley looked at giant squid axons. They used giant squids because squids had axons large enough to manipulate and use their specially built glass electrodes on. (Click here for more information on materials and methods). From their experimentation with a squid axon, they were able to create a circuit model that seemed to match how the squid axon carried an action potential.

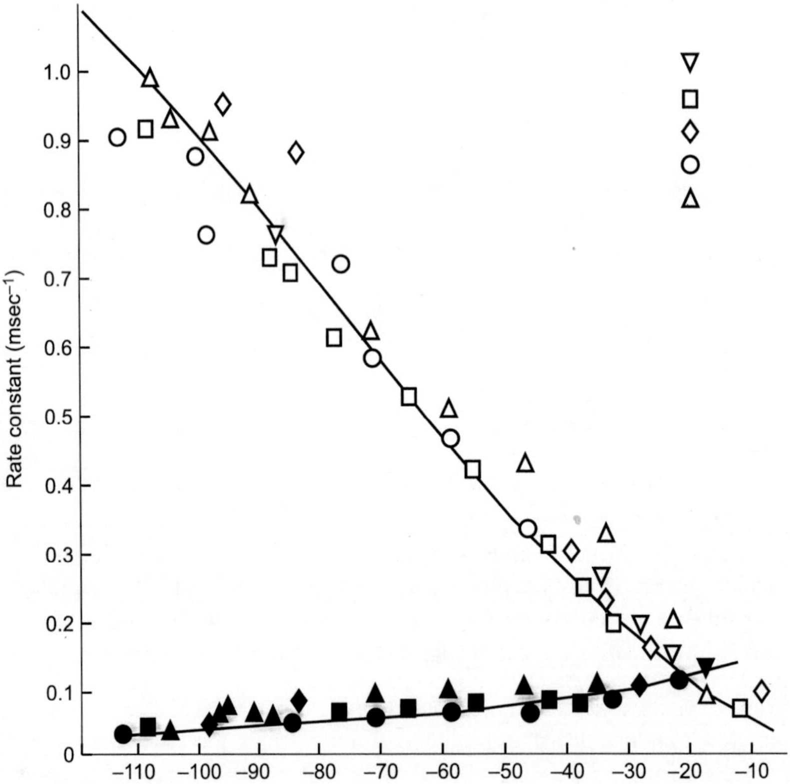

Current flowing through the membrane can be carried via the charging and discharging of a capacitor or via ions flowing through variable resistances in parallel with the capacitor. Each of the resistances corresponds to charge being carried by different components. In the nerve cell these components are sodium and potassium ions and a small leakage current that is associated with the movement of other ions, including calcium. Each current (I_{Na}, I_K, and I_L) can be determined by a driving force which is represented by a voltage difference and a permeability coefficient, which is represented by a conductance in the circuit diagram. Conductance is the inverse of resistance. These equations can easily be derived using Ohm's law ($V = IR$)

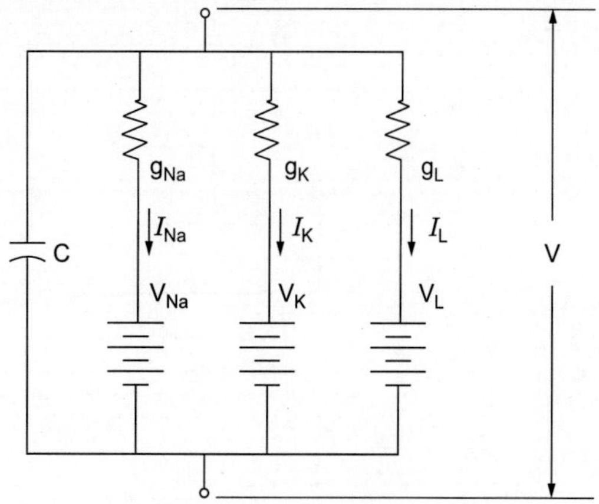

$$I_{Na} = g_{Na}(E - E_{Na})$$
$$I_K = g_K(E - E_K)$$
$$I_L = g_L(E - E_L)$$

g_{NA} and g_K are both functions of time and membrane potential. E_{Na}, E_K, E_L, C_m and g_L are all constants that are determined via experimentation. (Click here for more information on currents).

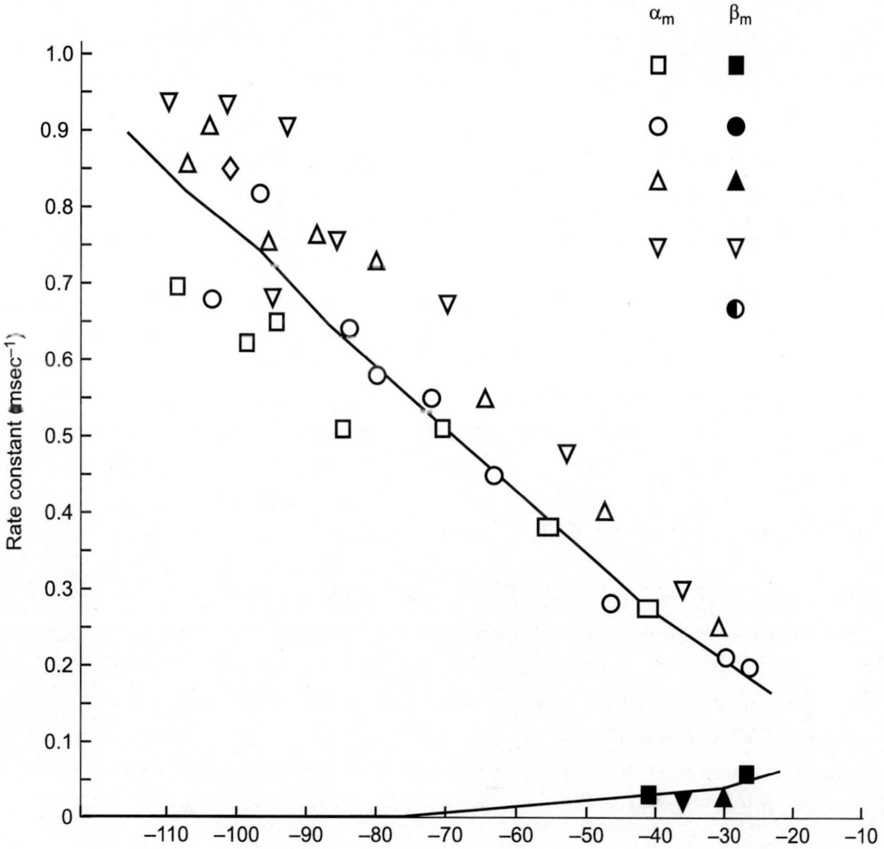

Graph used to determine the values for the conductance rate constants (m)

The influences of membrane potential on permeability were discovered to perform as follows. Under depolarization conditions, there is a transient increase in sodium conductance and a slower but more sustained increase in potassium conductance. These changes can be reversed during repolarization. The nature of these permeability changes was not fully understood when Hodgkin and Huxley did their

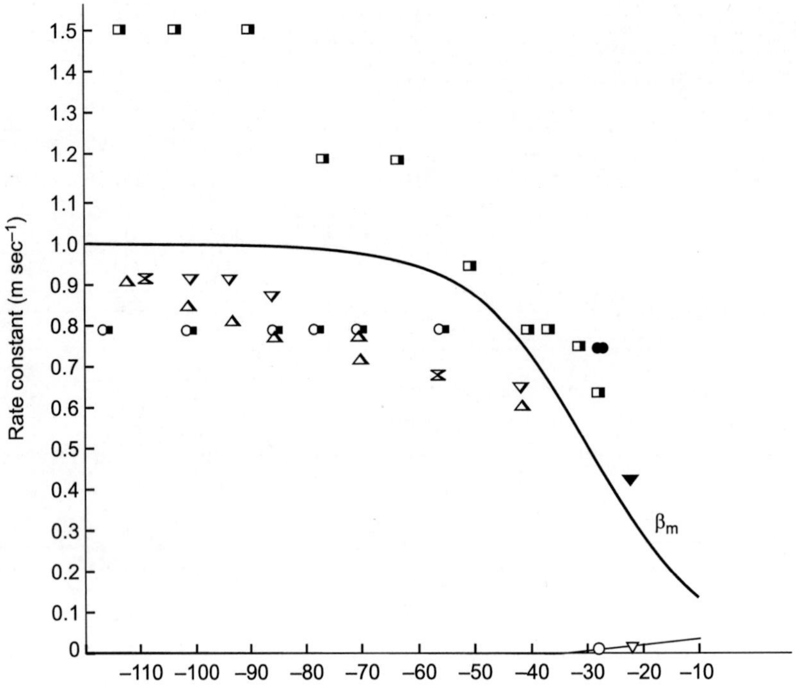

Graph used to determine the values for the conductance rate constants (h), a

work. They did not know what the cellular membrane looked like on the micro scale. They did not know about the existence of ion channels and ion pumps in the membrane. Based off of their finding, however, they were able to conclude that changes in permeability were dependant on membrane potential and not membrane current. Molecules aligning or moving with the electric field cause a change in permeability. Originally they supposed that sodium ions crossed the membrane via lipid carrier molecules that were negatively charged. What they observed however, proved that this was not the case. Rather, they supposed that sodium movement depends on the distribution of charged particles which do not act as carriers in the usual sense but rather allow sodium to pass through the membrane when they occupy particular sites on the membrane. This turned out to be the case. These charged particles are ion channels. In the case of sodium permeability, the carrier molecules (as they are referred to by Hodgkin and Huxley) are inactivated when there is a high potential difference. Potassium permeability is similar to sodium permeability but there are some key differences. The activating carrier molecules have an affinity for potassium, not sodium. They move more slowly and they are not blocked or inactivated. (Click here for more information on conductances)

To build their mathematical model that describes how the membrane current works during the voltage clamp experiment, they used the basic circuit equation

$$I = C_m \frac{dV}{dt} + I_i$$

where I is the total membrane current density (inward current positive). It is the ionic current density (inward current positive), V is the displacement of membrane potential (depolarization is negative), C_m is the membrane capacitance, t is time. They chose to model the capacity current and ionic current in parallel because they found that the ionic current when the derivative was set to zero and the capacity current when the ionic current is set to zero were similar. We can enrich this equation further by realizing that

$$I_i = I_{Na} + I_K + I_L$$

where I_{Na} is the sodium current, I_K is the potassium current and I_L is the leakage current. We can further expand on this model by adding the following relationships:

$$I_{Na} = g_{Na}\,(V - V_{Na}) = g_{Na}(E_{Na} - E_R)$$
$$I_K = g_K(V - V_K) = g_K(E_K - E_R)$$
$$I_L = g_L(V - V_L) = g_L(E_L - E_R)$$
$$V = E - E_R$$

Where E_R is the resting potential. When examining the graph of the potassium current versus the potassium potential difference, you can see that in the beginning, it's or third order equation will describe it. But at the end, during the end, it seems to be more first order. In order to explain this in the conductance formula, we let

$$g_k = \overline{g_k} n^4$$

where $\overline{g_k}$ is a constant, and n is a dimensionless variable that varies from 0 to 1. It is the proportion of ion channels that are open. To further understand where n comes from we can derive the equation

$$\frac{dn}{dt} = \alpha_n(1 - n) - \beta_n n$$

where alpha is the rate of closing of the channels and beta is the rate of opening. Together, they give us the total rate of change in the channels during an action potential. The sodium conductance is described by the equation

$$g_{Na} = m^3 h \overline{g_{Na}}$$

where $\overline{g_{Na}}$ is a constant and m is the proportion of activating carrier molecules (ion channels) and h is the proportion of inactivation carrier molecules (ion channels). M and h can be further described by

$$\frac{dm}{dt} = \alpha_m(1 - m) - \beta_m m$$

$$\frac{dh}{dt} = \alpha_k(1 - h) - \beta_k h$$

where alpha and beta are again rate constants that are similar to the rate constants for the potassium conductance. (Click here for more information on inactivation).

APPENDIX 2: THE COHEN-GROSSBERG THEOREM

We close this chapter with a theorem proposed by Cohen and Grossberg [106] which has far reaching implications for neural network theory. They described a generalized model of a non-linear dynamical system and proved its global asymptotic stability by suggesting an appropriate Lyapunov function. It was later shown that a number of major neural network models were indeed special cases of this general system [195]. Below we present this theorem and a proof of its Lyapunov property.

Models that can be written in the form:

$$\frac{dx_i}{dt} = a_i(x_i)\left(b_i(x_i) - \sum_{j=1}^{n} c_{ji} d_j(x_j) \right) \quad i = 1, \ldots, n \qquad \ldots(1)$$

admit the global Lyapunov function

$$E = -\sum_{i=1}^{n} \int_0^{x_i} b_i(\alpha_i)\, d_i'(\alpha_i) d\alpha_i + \frac{1}{2} \sum_{j,k=1}^{n} c_{jk} d_j d_k \qquad \ldots(2)$$

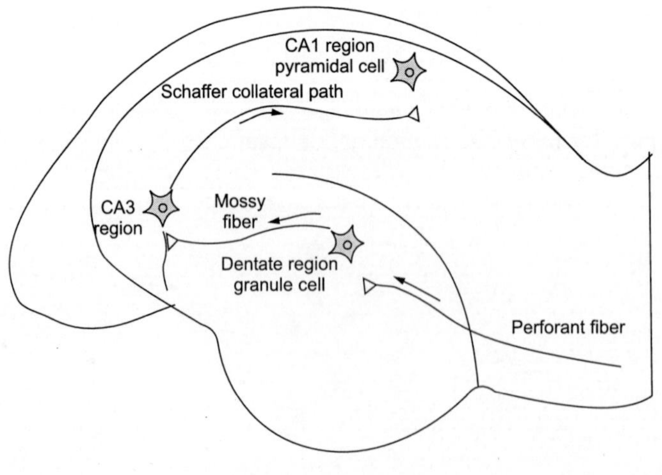

Fig. A2.1 *Hippocampal pathways believed to be involved in long-term potentiation*

It is most interesting that CA1 LTP displays three important properties: cooperativity, associativity, and specificity. Cooperativity implies that CA1 LTP cannot be produced by activating only one fiber. There is a minimum number of fibers that must be activated together to achieve LTP. This cooperativity has an associative feature: when both strong and weak excitatory inputs arrive in the same dendritic region of a pyramidal cell, the weak input gets potentiated if it is activated in association with the strong one. Finally, LTP is specific to the dendrites where it is produced.

Amongst various other observations, the most important point to note is that LTP requires depolarization of the postsynaptic neuron, and that this depolarization must coincide with fhe stimulus from the presynaptic neuron. This is in fact the first direct evidence for the Hebbian Learning Rule first proposed by Canadian psychologist Donald Hebb [222]:

> *"When an axon of cell A . . . excite[s] cell B and repeatedly or persistently takes part in firing it, some growth process or metabolic change takes place in one or both cells so that A's efficacy as one of the cells firing B is increased."*

The importance and ramifications of Hebb's postulate can hardly be overemphasized. Later in this chapter, and in those to come, we shall see that this principle underlies a large number of neural network models. It motivates learning laws that are employed in these models, both directly

When A_i and B_i are in different spaces, $A_i \in \mathbb{B}^n$, $B_i \in \mathbb{B}^m$ then the model is a *heteroassociative memory* which associates two different vectors with one another. If $B_i = A_i \in \mathbb{B}^n$ then the model is an *autoassociative memory* which associates a vector with itself. Associative memory models properties such as fault tolerance and, when they are autoassociative, they are able to solve the pattern completion problem.

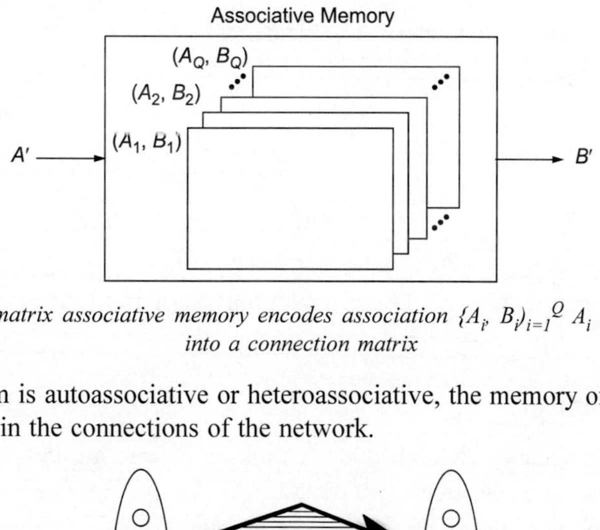

Fig. A2.2 *A matrix associative memory encodes association $\{A_i, B_i\}_{i=1}^{Q}$ $A_i \in \mathbb{B}^n$, $B_i \in \mathbb{B}^m$ into a connection matrix*

Whether the system is autoassociative or heteroassociative, the memory of the vectors that are to be associated is stored in the connections of the network.

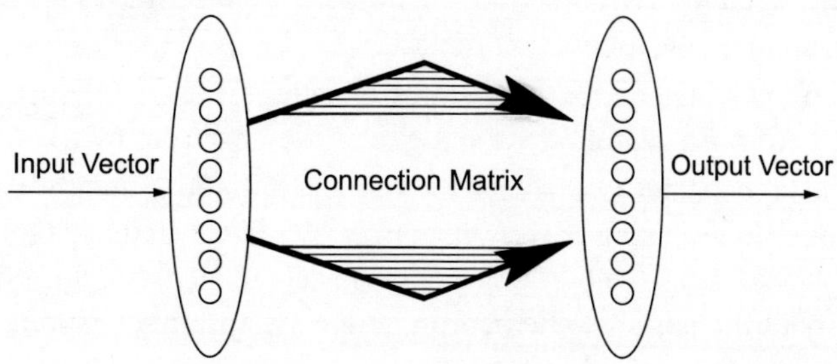

Fig. A2.3 *A simple associative memory is generated by a two layer neural network.*

Figure shows two layers or fields of neurons, one which receives input vectors and one which generates output vectors. The two layers are assumed to be connected through an $n \times m$ connection matrix **W**. As usual, neurons integrate activations linearly. If all the neurons are assumed linear, then for an input vector A that is presented to the network, the activation and thus the signal vector B of output neurons can be written compactly in matrix form:

$$B = A^T\mathbf{W} \quad \text{...(3)}$$

Here, the association of B with A is effected through the connection matrix **W**. Such a network is purely linear.

Additional complexity can be introduced into the operation of the same network by employing non-linear signal functions at the input as well as at the output. In this case, we would write Eq. (3) as:

$$B = \delta(A^T\mathbf{W}) \quad \text{...(4)}$$

where we assume that the signal vector A is generated using a non-linear function of activations in the input layer. In this case the memory is non-linear.

Note that the network architecture portrayed in Fig. allows activations to flow unidirectionally from the input layer to the output layer of neurons. A further level of complexity in the operation of the same network can be introduced by incorporating feedback. As shown in Fig. an additional set of connections allows signals from the output layer of neurons to flow back to the input layer, thereby modifying their activation levels. This introduces dynamnics into the network operation.

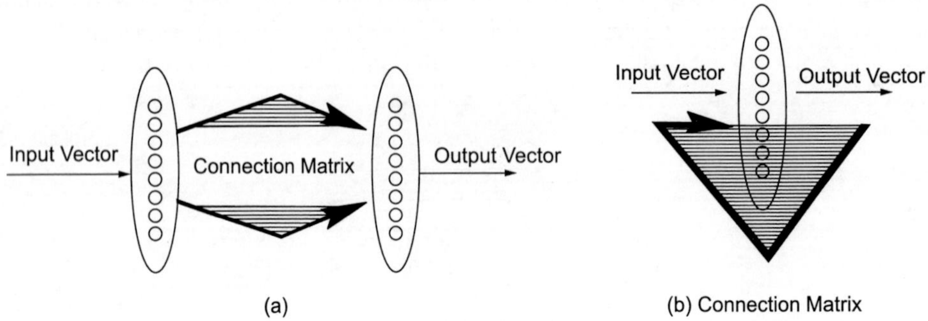

Fig. A2.4 *Associative memories generally employ feedback*

In Fig. a single layer of neurons feeds back to itself through a single set of connections generating an autoassociative architecture. We will have an opportunity to study all these variants of neural network models.

Hebb Association Matrix

Importantly, since the association is embedded within the connection matrix, we need address the question: How do we compute the connection matrix?

Recall that the Hebb postulate essentially states that the synaptic efficacy is proportional to the conjunction (or product) of postsynaptic neuron signals. We construct matrix **W** in accordance low.

To formalize this idea, consider two layers F_X and F_Y of n and m neurons respectively, such that neurons from F_X connect to F_Y using a weight matrix **W**. If we assume that the weight matrix is constructed using Hebbian learning, it modifies synapses in accordance with the product of the presynaptic activity with the activity of the postsynaptic neuron, at the synapse in question:

$$\delta w_{ij} = \eta a_i b_j \quad \text{...(5)}$$

for some learning rate η, and signal vectors $A = (a_1, \ldots, a_n)^T$ and $B = (b_1, \ldots, b_m)^T$ across F_X and F_Y respectively. This rule is often referred as the *generalized Hebb rule.* If we assume the learning rate $\eta = 1$, Hebbian learning essentially reduces to taking the outer products of signal vectors. For example, if we start out with a zero association matrix, then a single association between two vectors A and B is then encoded as:

$$\mathbf{W} = \begin{pmatrix} a_1b_1 & \cdots & a_1b_m \\ \vdots & \ddots & \vdots \\ a_nb_1 & \cdots & a_nb_m \end{pmatrix} = AB^T \qquad \ldots(6)$$

The Hebb conjunctional learning rule should be clear from the above matrix: the weight w_{ij} that connects neuron i of F_X to neuron j of F_Y is the product of the respective signals a_i and b_j of both the neurons. We then say that the association (A, B) has been encoded into the matrix. We will revisit this form of conjunctional learning in the context of all the attractor neural networks we study in this chapter.

Linear Associative Memory

It may come as a surprise that a connection matrix that is formed by taking the outer product of two vectors as in Eq. (6) can make a simple associative memory [13]. To keep the discussion simple, we work with linear neurons. We assume that a single association (A, B) has been encoded into the connection matrix $\mathbf{W}$, using the outer product: $\mathbf{W} = AB^T$. If we present the vector A to F_X, then the activity that develops across F_Y is:

$$(B')^T = A^T\mathbf{W} = A^TAB^T = ||A||^2B^T \qquad \ldots(7)$$

The output vector that is *recalled* across the F_Y is a scaled version of B, the associant of A. If the vector A is normalized, recall is exact.

Now what if we wish to encode more than one association? The natural course to follow is to superpose the individual association matrices.

Assuming Q associations $\{A_k, B_k\}_{k=1}^{Q}$ that are to be encoded, the resulting connection matrix is:

$$\mathbf{W} = \sum_{k=1}^{Q} A_k B_k^T \qquad \ldots(8)$$

which is the sum of outer products of individual associations. Of course such a simple superposition will certainly cause the individual memories to interfere leading to pathological behaviour of the model. We will return to this issue in sections that follow.

A special (and very restrictive) case arises if the vectors A_k are *orthonormal*—orthogonal to one another and normalized to unity magnitude. In such a case, when we present a single association A_i to neurons in field F_X, a forward pass through the network yields a vector B':

$$(B')^T = A_i^T\mathbf{W} = A_i^T\sum_{k=1}^{Q} A_k B_k^T \qquad \ldots(9)$$

$$= A_i^T A_i B_i^T + \sum_{k\neq 1} A_i^T A_k B_k^T \qquad \ldots(10)$$

$$= \|A_i\|^2 B_i^T + 0 = B_i^T \qquad \ldots(11)$$

The recall of the encoded associant B_i is perfect.

Finally, what happens if we present vectors that are close to the encoded associations? The following example provides insight.

Orthogonal Linear One-shot Associative Memories

Consider encoding the following three associations into a linear memory:

$$A_1 = (1 \;\; 0 \;\; 0 \;\; 0)^T \quad B_1 = (1 \;\; 2 \;\; 3)^T \qquad \text{...(12)}$$

$$A_2 = (0 \;\; 1 \;\; 0 \;\; 0)^T \quad B_2 = (-2 \;\; 3 \;\; 1)^T \qquad \text{...(13)}$$

$$A_3 = (0 \;\; 0 \;\; 1 \;\; 0)^T \quad B_3 = (4 \;\; 0 \;\; 4)^T \qquad \text{...(14)}$$

where A_1, A_2, A_3, B_1, B_2, B_3 are column vector associations. Encoding generates the following connection matrix:

$$\mathbf{W} = A_1B_1^T + A_2B_2^T + A_3B_3^T \qquad \text{...(15)}$$

$$= \begin{pmatrix} 1 & 2 & 3 \\ 0 & 0 & 0 \\ 0 & 0 & 0 \\ 0 & 0 & 0 \end{pmatrix} + \begin{pmatrix} 0 & 0 & 0 \\ -2 & 3 & 1 \\ 0 & 0 & 0 \\ 0 & 0 & 0 \end{pmatrix} + \begin{pmatrix} 0 & 0 & 0 \\ 0 & 0 & 0 \\ 4 & 0 & 4 \\ 0 & 0 & 0 \end{pmatrix} = \begin{pmatrix} 1 & 2 & 3 \\ -2 & 3 & 1 \\ 4 & 0 & 4 \\ 0 & 0 & 0 \end{pmatrix} \qquad \text{...(16)}$$

APPENDIX 3: PERCEPTRON LEARNING RULE CONVERGENCE THEOREM

Perceptron Convergence Theorem: Says that there if there is a vector w^* such that $f(w^* \, p(q)) = t(q)$ for all q, then for any starting vector w, the perceptron learning rule will converge to a weight vector (not necessarily unique and not necessarily w*) that gives the correct response for all training patterns, and it will do so in a finite number of steps.

Idea of the Proof: The idea is to find upper and lower bounds on the length of the weight vector. If the length is finite, then the perceptron has converged, which also implies that the weights have changed a finite number of times.

Proof:

1. Assume that the inputs to the perceptron originate from two linearly separable classes. **That is, the classes can be distinguished by a perceptron**. Let X_1 be the subset of training vectors belonging to C_1. That is, $p(1), p(2), \ldots$. Let X_2 be the set of training vectors belonging to C_2. That is, $p(1), p(2), \ldots$. Then, we can say that $X_1 \cap X_2$ is the complete training set X.
2. Given the set of vectors X_1 and X_2 to train this perceptron to train this perceptron, the training process (as we have seen) involves the adjustment of the weight vector **w** such that C_1 and C_2 are linearly separable. That is, there exists some **w** such that

$$\mathbf{w}^T\mathbf{p} > 0 \text{ for every input vector } \mathbf{p} \in C_1$$
$$\mathbf{w}^T\mathbf{p} < 0 \text{ for every input vector } \mathbf{p} \in C_2 \quad \ldots(1)$$

3. What need to do is find some **w** such that the above is satisfied, which is the purpose of the perceptron algorithm.

   One algorithm for adapting the weight vector for the perceptron algorithm can be formulated as follows (there are others):

   (a) If the kth member of the training set $\mathbf{p}(k)$ is correctly classified at the kth iteration no correction is made to **w**. This is done according to the following rule:

$$\mathbf{w}(k+1) = \mathbf{w}(k) \text{ if } \mathbf{w}^T\mathbf{p}(k) > 0 \text{ and } \mathbf{p}(k) \in C_1$$
$$\mathbf{w}(k+1) = \mathbf{w}(k) \text{ if } \mathbf{w}^T\mathbf{p}(k) < 0 \text{ and } \mathbf{p}(k) \in C_2 \quad \ldots(2)$$

   (b) Otherwise, the weight vector of the perceptron is updated according to the following rule:

$$\mathbf{w}(k+1) = \mathbf{w}(k) - \mathbf{p}(k) \text{ if } \mathbf{w}^T\mathbf{p}(k) > 0 \text{ and } \mathbf{p}(k) \in C_2$$
$$\mathbf{w}(k+1) = \mathbf{w}(k) + \mathbf{p}(k) \text{ if } \mathbf{w}^T\mathbf{p}(k) < 0 \text{ and } \mathbf{p}(k) \in C_1 \quad \ldots(3)$$

4. The proof begins assuming that $\mathbf{w}(1) = 0$ (i.e., the zero vector). Suppose that $\mathbf{w}^T(k)\mathbf{p}(k) < 0$ for $k = 1, 2, ...,$ and all the input vectors $\mathbf{p}(k) \in X_1$ (or C_1).

   Here, the perceptron **incorrectly** classifies the vectors $\mathbf{p}(1), \mathbf{p}(2),...$ since the second condition in Equation (1) is violated (i.e., $\mathbf{w}^T(k)\mathbf{p}(k)$ should be greater than (0). Now we can use Equation (3) to write the adjustments to the weight matrix according to the perceptron learning rule

$$\mathbf{w}(k+1) = \mathbf{w}(k) + \mathbf{p}(k) \text{ for } \mathbf{p}(k) \in C_1 \qquad ...(4)$$

   Recall, what we are doing is modifying the weight vector so that it points in the right direction.

5. Given the initial condition $\mathbf{w}(1) = 0$, we can iteratively solve for $\mathbf{w}(k+1)$ obtaining:

$$\mathbf{w}(k+1) = \mathbf{p}(1) + \mathbf{p}(2) + ... + \mathbf{p}(\mathrm{k}) \qquad ...(5)$$

   CLS Exercise: Show $\mathbf{w}(k+1) = \mathbf{p}(1) + \mathbf{p}(2) + ... + \mathbf{p}(k)$

   Answers:

   $\mathbf{w}(1) = 0$

   $\mathbf{w}(2) = \mathbf{w}(1) + \mathbf{p}(1) = \mathbf{p}(1)$

   $\mathbf{w}(3) = \mathbf{w}(2) + \mathbf{p}(2) = \mathbf{w}(1) + \mathbf{p}(1) + \mathbf{p}(2) = \mathbf{p}(1) + \mathbf{p}(2)$

   $\mathbf{w}(4) = \mathbf{w}(3) + \mathbf{p}(3) = \mathbf{w}(2) + \mathbf{p}(2) + \mathbf{p}(3) = \mathbf{w}(1) + \mathbf{p}(1) + \mathbf{p}(2) + \mathbf{p}(3)$

   $= \mathbf{p}(1) + \mathbf{p}(2) + \mathbf{p}(3)$

   So, in general:

   $\mathbf{w}(k+1) = \mathbf{p}(1) + \mathbf{p}(2) + ... + \mathbf{p}(k)$

   Since C_1 and C_2 are assumed to be linearly separable, then there exists a $\mathbf{w}_0$ that can correctly classify all input vectors belonging to C_1 and C_2.

   In this proof, we have assumed the existence of $\mathbf{w}_0$ for which

$$\mathbf{w}_o^T\mathbf{p}(k) > 0 \text{ if } \mathbf{p}(k) \in C_1$$

   and
$$\mathbf{w}_o^T\mathbf{p}(k) < 0 \text{ if } \mathbf{p}(k) \in C_2$$

   This is equivalent to the existence of a weight vector $\mathbf{w}_o$ for which

$$\mathbf{w}_o^T\mathbf{p}(k) > 0 \text{ if } \mathbf{p}(k) \in C_1 \cup C_2 \text{ or } X$$

   The reason is that the training set can be considered to consist of two parts.

$$C_1 = \{\mathbf{p} \text{ such that the target value is } 1\}$$

   and
$$C_2 = \{\mathbf{p} \text{ such that the target value is } 0\}$$

   So we can think of the training set X as

$$X = C_1 \cup C_2$$

   where $C_2 = \{-\mathbf{p} \text{ such that } \mathbf{p} \in C_2\}$

   Now, if the response for the network is incorrect, then this allows the weights to be updated according to:

$$\mathbf{w}(k+1) = \mathbf{w}(k) + \mathbf{p}(k)$$

   since $-\mathbf{w}_o^T\mathbf{p}(k) < 0 \Rightarrow \mathbf{w}_o^T\mathbf{p}(k) > 0$ if $\mathbf{p}(k) \in C_2$. Now in this case the perceptron is updated by $\mathbf{w}(k+1) = \mathbf{w}(k) - (-\mathbf{p}(k)) = \mathbf{w}(k+1) = \mathbf{w}(k) + \mathbf{p}(k)$ since $\mathbf{w}^T\mathbf{p}(k) > 0$ and $\mathbf{p}(k) \in C_2$.

   For the solution $\mathbf{w}_o$, we can define some $\alpha > 0$ as

$$\alpha = \min_{\mathbf{p}(k) \in X} \mathbf{w}_o^T\mathbf{p}(k) \qquad ...(6)$$

   which is just the minimum (scalar) value of all $\mathbf{w}_o^T\mathbf{p}(k)$ for all $\mathbf{p} \in X$.

6. By multiplying each side of Equation (5) by $\mathbf{w}_o^T$, we get

$$\mathbf{w}_o^T\mathbf{w}(k+1) = \mathbf{w}_o^T\mathbf{p}(1) + \mathbf{w}_o^T\mathbf{p}(2) + \ldots + \mathbf{w}_o^T\mathbf{p}(k)$$

7. And, by applying Equation (6), we get

$$\mathbf{w}_o^T\mathbf{w}(k+1) \geq k\alpha \qquad \ldots(7)$$

since $\mathbf{w}_o^T\mathbf{w}(k+1)$ has to be greater than or equal to $k \times \min_{\mathbf{p}(k)\in X} \mathbf{w}_o^T\mathbf{p}(k)$.

8. Next, use the Cauchy-Swartz inequality for $\mathbf{w}_o^T$ and $\mathbf{w}_o^T(k+1)$, which states for any two vectors **x** and **y**:

$$[\mathbf{x}\cdot\mathbf{y}]^2 \leq ||\mathbf{x}||^2\ ||\mathbf{y}||^2 \quad \text{or} \quad ||\mathbf{x}||^2 \geq [\mathbf{x}\cdot\mathbf{y}]^2/||\mathbf{y}||^2$$

where $||\mathbf{x}||$ is the Euclidian norm.

Example. Suppose $\mathbf{x} = [-1\ 0\ 4]$. Then, $||\mathbf{x}|| = \sqrt{(-r) + 0^2 + 4^2} = \sqrt{17}$.

Thus, we can say:

$$\left\|\mathbf{w}_o^T\right\|^2 \left\|\mathbf{w}(k+1)\right\|^2 = [\mathbf{w}_o^T\mathbf{w}(k+1)]^2 \qquad \ldots(8)$$

9. From Equation (7), we know by applying Cauchy-Swartz that

$$\mathbf{w}_o^T\mathbf{w}(k+1) \geq k\alpha$$

$$\left\|\mathbf{w}_o^T\right\|^2 \left\|\mathbf{w}(k+1)\right\|^2 \geq [k\alpha]^2$$

$$\left\|\mathbf{w}_o^T\right\|^2 \left\|\mathbf{w}(k+1)\right\|^2 \geq k^2\alpha^2$$

and we get:

$$\left\|\mathbf{w}(k+1)\right\|^2 \geq \frac{k^2\alpha^2}{\left\|\mathbf{w}_o\right\|^2} \qquad \ldots(9)$$

which shows that the squared length of the weight vector ($||\mathbf{w}(k+1)||^2$) grows by a factor of k^2, where k is the number of time the weights have changed.

10. So, what we have established from Equation (9) is a **lower-bound** in the terms of the squared Euclidian norm of the **weight vector w at iteration** $k + 1$.

But there is another aspect we must consider. In order to show that the weights cannot continue to grow indefinitely, we must establish an **upper-bound** for the weight vector **w**.

11. To find an **upper-bound**, we now do the following. Write Equation (4) as the following (where q is the number of patterns in X):

$$\mathbf{w}(k+1) = \mathbf{w}(k) + \mathbf{p}(k) \text{ for } k = 1, \ldots, q\ \mathbf{p}(k) \in X$$

12. After taking the square of the Euclidian norm, we get

$$||\mathbf{w}(k+1)||^2 = ||\mathbf{w}(k)||^2 + 2\mathbf{w}^T(k)\ \mathbf{p}(k) + ||\mathbf{p}(k)||^2 \qquad \ldots(10)$$

13. But assuming that the perceptron incorrectly classifies an input vector belonging to X(*i.e.*, we have to make the adjustment $\mathbf{w}(k+1) = \mathbf{w}(k) + \mathbf{p}(k)$ since $\mathbf{w}^T(k)\ \mathbf{p}(k) < 0$, then from Equation (10), it follows that:

$$||\mathbf{w}(k+1)||^2 \leq ||\mathbf{w}(k)||^2 + ||\mathbf{p}(k)||^2$$

14. Observe that $\|\mathbf{w}(k+1)\|^2 \leq k\beta$ conflicts with the earlier result of $\|\mathbf{w}(k+1)\|^2 \geq \dfrac{k^2\alpha^2}{\|\mathbf{w}_o\|^2}$; namely, it can be shown that they don't agree for $\|\mathbf{w}(k+1)\|^2$ for large values of k.

    So we can state that k cannot be larger than some k_{max} for which Equations (9) and (14) are both satisfied; and they must be equal to determine the maximum number of iterations for two linearly separable classes to converge.

    To determine this, set Equations (9) and (14) equal to each other and substitute solve for k_{max}. Thus we get.

$$\frac{k_{max}^2\alpha^2}{\|\mathbf{w}_o\|^2} = k_{max}\beta$$

$$k_{max} = \frac{\|\mathbf{w}_o\|^2\beta}{\alpha^2}$$

    which means that changing the weights of the perceptron must terminate after at most k_{max} iterations, which means that the machine has solved the (linearly separable) problem correctly.

Spiking neurons 1

NCSI 850
Queen's University
Centre for Neuroscience Studies
Guest Lecture by Dominic Standage

Preamble

The brain is an organ that evolved to process information
i.e. the brain is a computer (or a sort)

What computational properties are conveyed by specific biophysical details?

There is no "best" level of analysis in neuroscience
Computational methods are often used to bridge levels of analysis

When are abstractions defensible, and when are they not?

Two trade-offs
(1) detail vs. computational efficiency
(2) detail vs. predictive power

A biological neuron

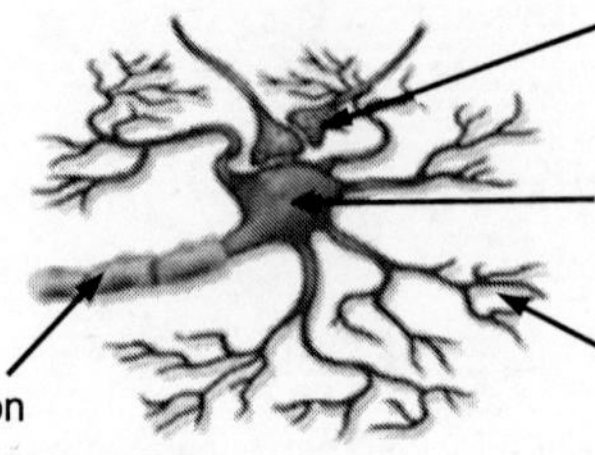

Wa're leaving out a lot at detail

Electrical circuit primer

Ion : an atom or group of atoms with non-zero net electric charge

i.e. # protons ≠ # electrons

Cations : excess positive charge, *e.g.* K+, Na+

Anions: excess negative charge, *e.g.* Cl–

Opposite charges attract : +–

Like charges repel: + + or – –

Electrical circuit primer

Ohm's law : V = 18

V : voltage or *potential difference*

I : current

R : resistance

or

G = 1/R : conductance

Electrical circuit primer

Ions have a statistical tendency to be uniformly distributed

Attractive and repulsive forces lead to movement

Current I is the movement of ions

Defined by convention as the direction of net positive charge

Cations move in the direction of current

Ohm's law : $I = V/R$

Electrical circuit primer

Resistance R is the opposition of a circuit to the flow of current imposed by the medium in which ions are making

Ohm's law : $R = V / I$

Conductance G is the inverse of R

Ohm's law : $G = I / V$

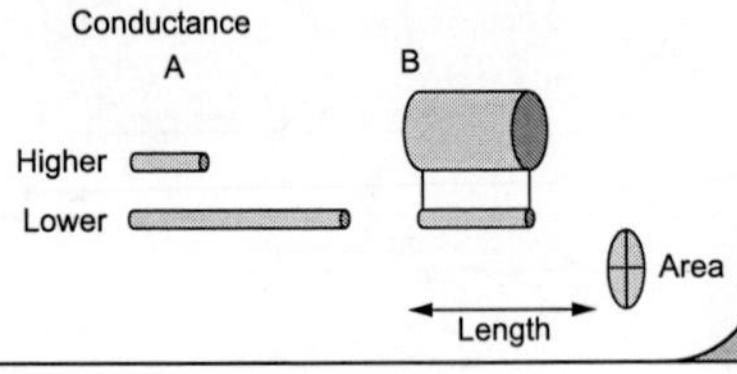

Electrical circuit primer

Voltage V measures potential difference

The energy required to move ions from one location to another

i.e. the potential for current to flow

Ohm's low : $V = I R$

Electrical circuit primer

Capacitance C stores charge Q across an insulator
Positive charges on one side, negative charges on the other
These is a potential difference across the capacitor

Resistive current tr cannot flow across the insulator
Charges are redistributed on either side
Capacitive current A/c

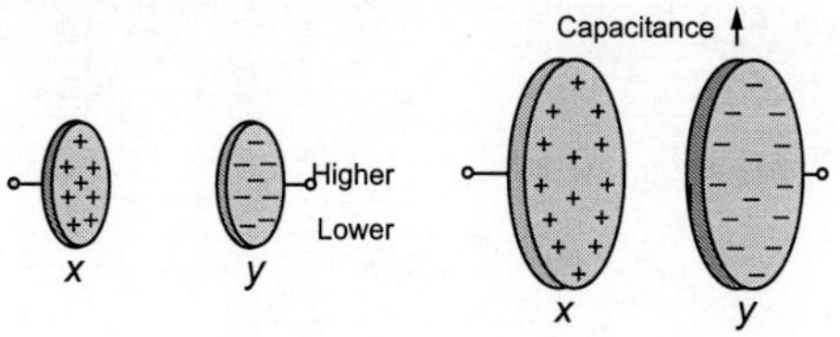

kendal, schwortz, and ressell, principles of Neurosciecn 2nd ed.

A neuron as resistor and capacitor in parallel

Ion channels

Source of current *Ir*

Source of charge for capacitor *C*

and thus current *IC*

im = *ir* + *iC*

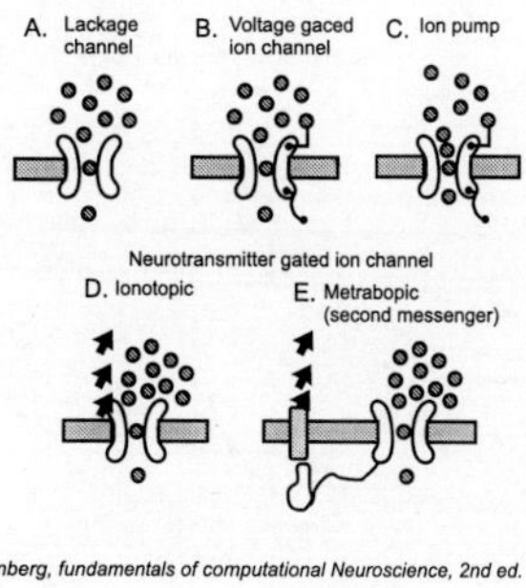

Tappanberg, fundamentals of computational Neuroscience, 2nd ed.

A neuron as resistor and capacitor in parallel

- Conductors/resistors representing ion channels
- Capacitor to store charge across the cell membrane

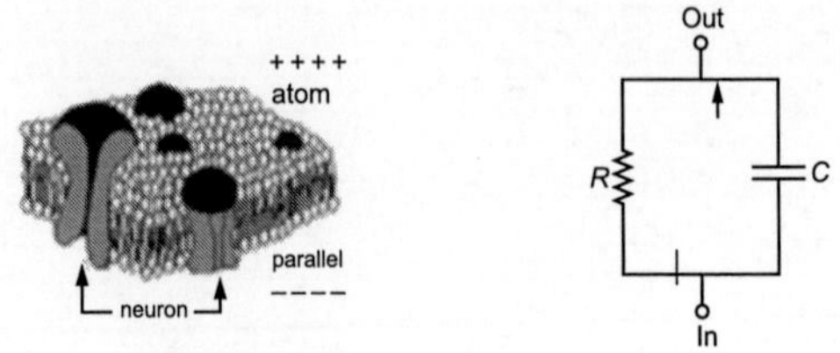

A neuron as resistor and capacitor in parallel

Together R and C determine the membrane time constant t

More conventions extracellular side is ground (0V)

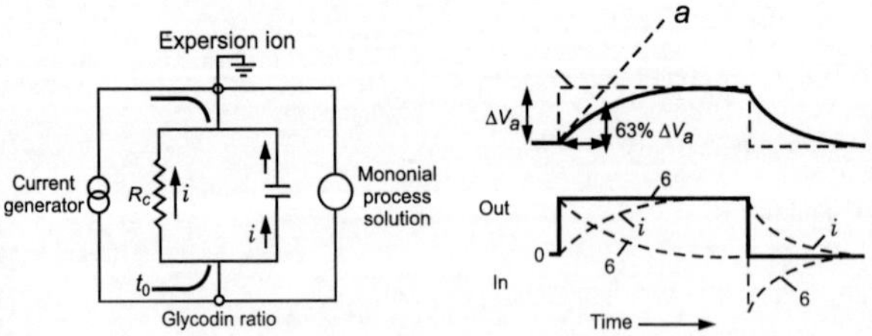

Equilibrium potential

Balance between a given concentration gradient and the potential difference it causes

i.e. the chemical and electrical driving forces and equal and opposite

ion pumps maintaion gradients

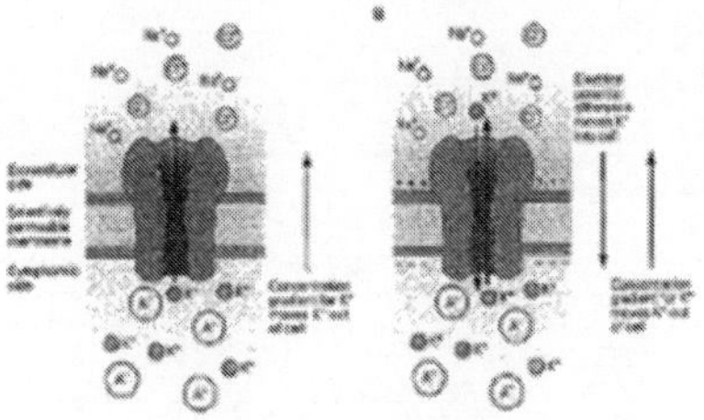

Nearst potential

Ex = RT/=F (Max)

Ex = 50mV/z log [Max]

e.g. 77mV for K+

Kandel, Schwartz and Jessell, Principles of Neural Science, 4th ed

Equilibrium potential

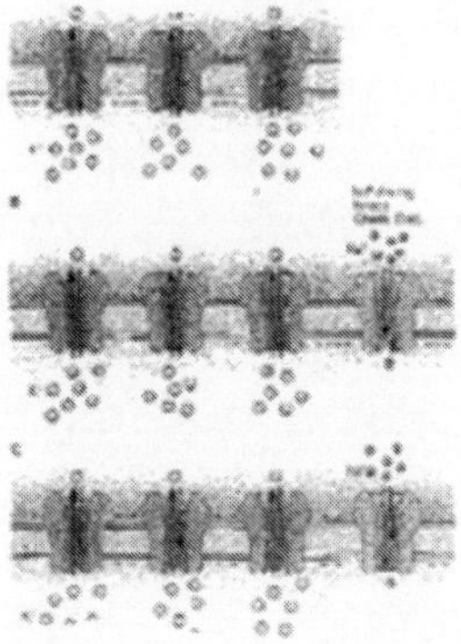

Equilibrium (ask. resting leakage) potential of the neuron is determined by the proparation of each type of ion channel

e.g. the resting potential is closer to the Nearnst potential of K+ (–77 mV) than Na+ (55 mV) because there are more K+ channel

$$I_m = \sum_t g_i(V - E_i)$$

– 60 - 70 mV for most neurons

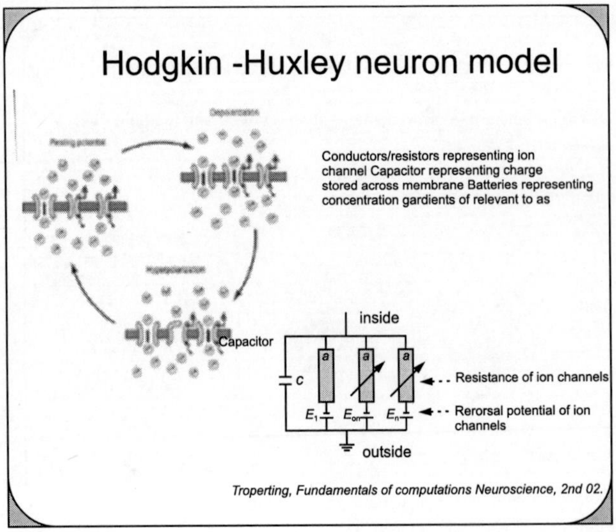

Hongkin -Huxley neuron model

$$I_j = g_j(V - E_j)$$ Ohm's law for each ion channel *j*

$$g_L(V - E_L)$$ Total passive/leakage current

$$g_{Na} = g_{Na} m^3 h$$

Active (voltage-dependent) currents for K+ and Na+

$$g_K = g_K n^4$$

$$\frac{dx}{dt} = -\frac{1}{\tau_x(V)} | x - x_0(V)] | \; x \in \{m, m, h\}$$

$$C\frac{dV}{dt} = -\sum_j g_j(V - E_j) + l_e(t)$$

Hodgkin -Huxley neuron model

k + conductance — Na+ — Leakage — Electrode

$$C\frac{dV}{dt} - - g_k n^4(V - E_k) - g_{Na}m^3h(V - E_{Na}) - g_L(V - E_L) + I(t)$$

Voltage-dependent dynamics

More conventions
Membrane current is defined as positive outward
Electrode currents are defined as positive inward

$$t_n(V)\frac{dn}{dt} = - [n - n_g(V)]$$ K^+ activation

$$t_m(V)\frac{dm}{dt} = - [m - m_o(V)]$$ Na^+ activation

$$t_h(V)\frac{dh}{dt} = - [h - h_g(V)]$$ Na^+ deactivation

Each time constant and × o(V) term were fit tha data

Easily numerically integrated e.g. Euler method

–or Matlab's ODE solvers

Hodgkin -Huxley neuron model

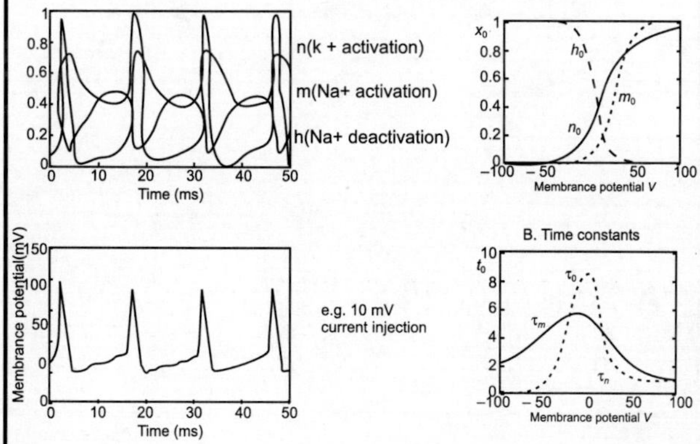

Transparting, fundamental of computational Neuroscience, 2nd ed.

Leaky integrate-and-fire neuron

If all spikes from a given neuron are the same, down need to model the spike?

Timing conveys information, not spikes them selves

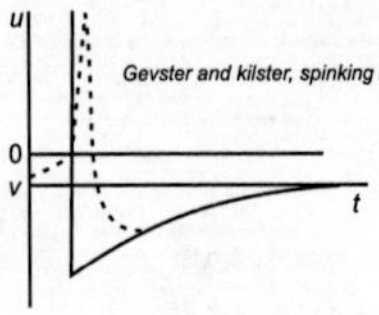

Please Provide

Sub-threshold dynamics only

AP generation mechanism can be modulated, so we can't always defend leaving out the spike – and other issues for later

Leaky integrate-and-fire neuron

Basic sub-threshold formulation of R and C in parallel

$$I = I_R + I_C$$

$$V = I_R R \ ; \ I_R = \frac{V}{R}$$

$$C = \frac{Q}{V}; \ CV = Q \ ; \ C\frac{dV}{dt} = \frac{dQ}{dt}; C\frac{dV}{dt} = I_c$$

$$I = \frac{V}{R} + C\frac{dV}{dt}; \ C\frac{dV}{dt} = - \ \frac{V}{R} + I$$

$$CR\frac{dV}{dt} = - \ V + IR \ ; \ t_m \frac{dV}{dt} = - \ V + IR$$

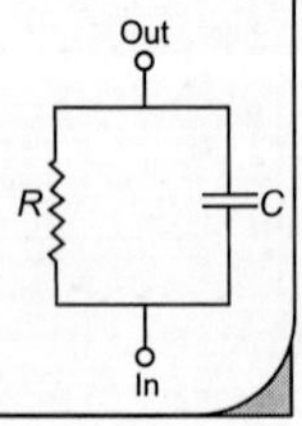

Leaky integrate-and-fire neuron

$$t_m \frac{dV}{dt} = -V + IR$$ — does not consider equilibrium potential

Reminder of conventions
membrance current defined as negative inward
Electrode current defined as positive inward

$$t_m \frac{dV}{dt} = (E_L - V) + RI_e$$

Leaky integrate-and-fire neuron

i_m Total current per unit area of cell

A Total surface area of cell

$$I_m = i_m A$$

$g_j(V - E_j)$ Conductance g
indes j for each type of channel
Current vcanishes for $V = E_j$

$$i_m = \sum_j g_j(V - E_j) \quad ; \quad i_m = g_L(V - E_L)$$

$$C\frac{dV}{dt} = I_m \quad ; \quad c_m = C/A \quad ; \quad C_m \frac{dV}{dt} = i_m$$

$$C_m \frac{dV}{dt} = i_m + \frac{I_e}{A}$$

Leaky integrate-and-fire neuron

$$c_m \frac{dV}{dt} = -i_m + \frac{I_e}{A} \quad ; \quad c_m \frac{dV}{dt} = -g_L(V - E_L) + \frac{I_e}{A}$$

$$r_m = \frac{1}{g_L}$$ Membrane resistance per unit area

$$\frac{1}{g_L} c_m \frac{dV}{dt} = \frac{1}{g_L} \times (-g_L(V - E_L) + \frac{I_e}{A})$$

$$r_m c_m \frac{dV}{dt} = -(V - E_L) + I_e \; ; \; \tau_m = c_m r_m$$

$$\tau_m \frac{dV}{dt} = E_L - V + RI_e$$

Leaky integrate-and-fire neuron

$$\tau_m \frac{dV}{dt} = E_L - V + RI_e$$ Subthreshold dynamic

$$t^f : V(t^f) = V_{thr}$$ Spike threshold

$$\lim_{t \to t^f \,;\, t > t^f} V(t) = V_{\text{res}}$$ Reset

$$V(t^f + \Delta^{abs}) = V_{\text{res}}$$ Absolute refractory period

Leaky integrate-and-fire neuron

$$\tau_m \frac{dV}{dt} = E_L - V + RI_e$$ LIF subthreshold

$$x(t + \Delta t) = x(t) + \Delta t f(x(t) \,.\, t)$$ Spike threshold

```
V = I ; F = I;
dt = 1%; time tep (μr)
for t = 1 : 100% ms runing time
v v + dt/tk% m* (E_L – V + R*) {t – t_F > tau _ ref}
f = v > v _ th boolean
% reset time to v_ res
v = v + f* (–v + v_res)
t_f = f* t + (1 – f) t % for abs refrac period
% Record firing
if f
F = (F = t); end
V = [v; v];
end
```

Leaky integrate-and-fire neuron

$$\tau_m \frac{dV}{dt} = E_L - V + RI_e$$ LIF subthreshold

$$x(t + \Delta t) = x(t) + \Delta t f(x(t) \,.\, t)$$ Euler integration

A. External input R/ωt = 8 mV < threshold B. External input Rf_{ext} = 12 mV > threshold

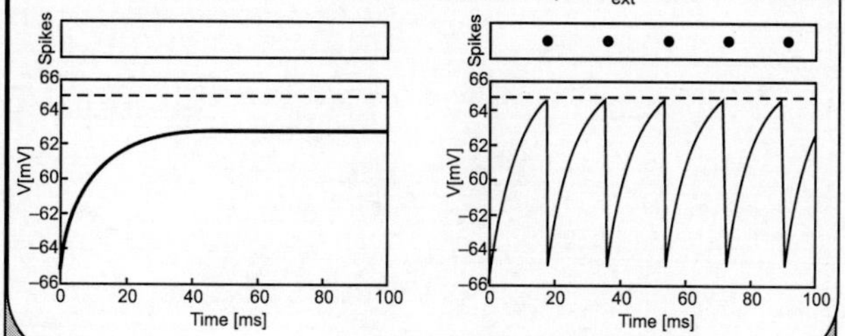

Transparting, fundamental of computational Neuroscience, 2nd ed.

Leaky integrate-and-fire neuron

LIF is computationally cheap, but has many limitations

No response adptation, rebound potential, bursting, etc..

– subthreshold potential is concave down. Could that matter?

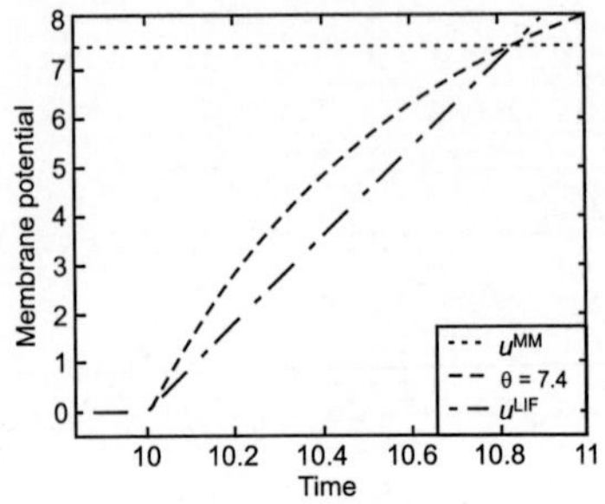

Modelling synapses

With point neurons, we want to simulate responses to spiking activity

aka.Spike Response Models

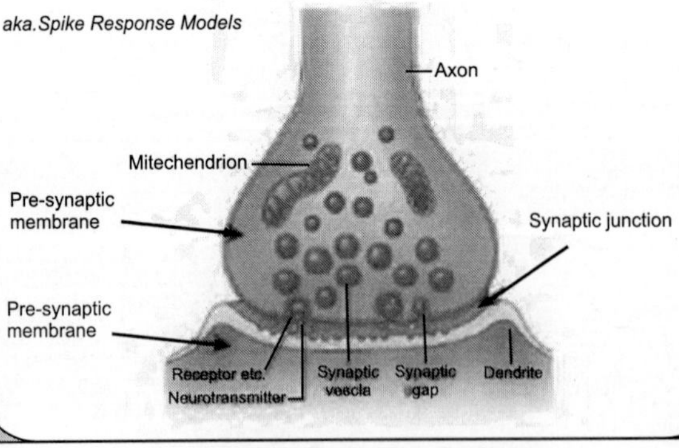

Modelling synapses

$$I_j = p\, a_j\, g_j\, (V - E_j)\, \mathrm{W}\, \eta$$

Typical values
Excitatory synapses = Ej = 0 mV
Inhibitory synapses: ej = .70 mV

I I I III II I Input spike train

$$\frac{da_j}{dt} = -\frac{a_j}{T_j} + \mathrm{d}(t - t_f)$$

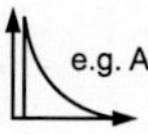

e.g. AMPA and GABA

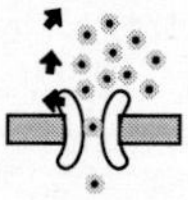

$$\frac{d_{aNMDA}}{dt} = -\frac{a_{NMDA}}{\mathrm{t}_{NMDA}} + rx(1 - a_{NMDA}) \;;\; \frac{dx}{dt} = -\frac{x}{\mathrm{t}_x} + \mathrm{d}(t - t_f)$$

$\eta = 1/(1 + \mathrm{Mg} \cdot \exp(-0.062.\mathrm{V}/3.57))$

Jahr and strene/Neurosci, 1990

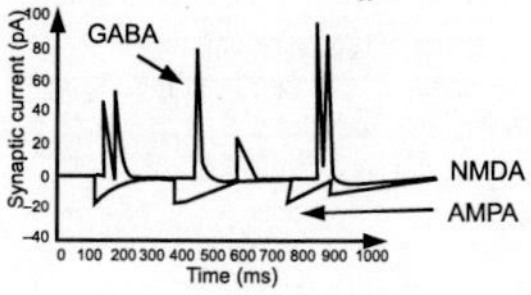

Modelling synapses

$$\tau_m \frac{dV}{dt} = E_L - V - RI_m + RI_e$$

Include synaptic current ...

... defined as, e.g.

$$I_m = a_A\, g_A\, (V - E_A) \quad - \quad a_N\, g_N\, (V - E_N)\, \eta \quad - \quad a_G\, g_{NMDA}(V - E_G)$$

$$\frac{da_j}{dt} = -\frac{a_j}{T_j} + \delta(t - t_f)$$

I I I III II I

$$\frac{d_{aNMDA}}{dt} = -\frac{a_{NMDA}}{\tau_{NMDA}} + rx(1 - a_{NMDA}); \quad \frac{x}{\tau_x} \quad + \delta(t - t_f)$$

Izhikevich neuron

HH neuron : fairly thorough characterization of neural dynamics
... but still a point neuron
LIF neuron : computationally

Is there anything in between ?

e.g. the Izhikevich neuron model
Very flexible, can simulate a variety of spiking behaviour, and computationally cheamp

$$v' = 0.004^2 + 5v + 140 - u + I \qquad \text{if } v \geq 30 \text{ mV, then} \begin{cases} u \leftarrow c \\ u \leftarrow u + d \end{cases}$$
$$u' = a(bc - u)$$

a : timescale of recovery variable u
b : sensitively of u to subthreshold fluctuations of v ; how strongly u and v are coupled
c : after spikae reset of v
d: after spicka recovery of u

E. Izhikevich, IEEE Trans, in Neurol Networks, 16:2003

Izhikevich neuron

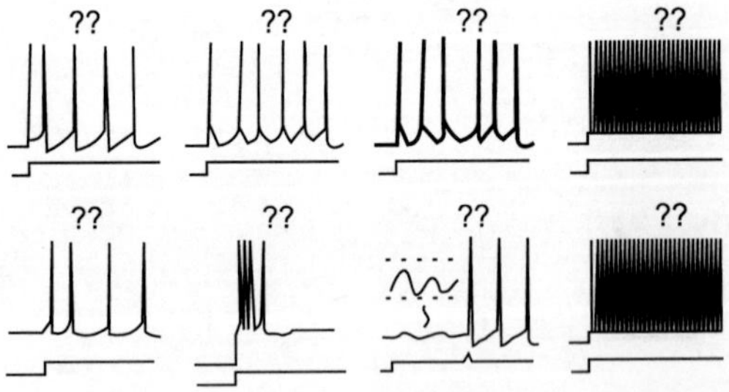

E. Izhikevich, IEEE Trans, in Neurol Networks, 16:2003

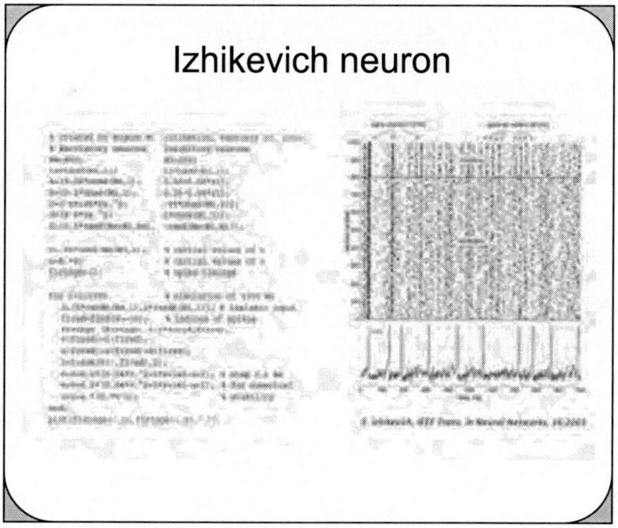
Izhikevich neuron

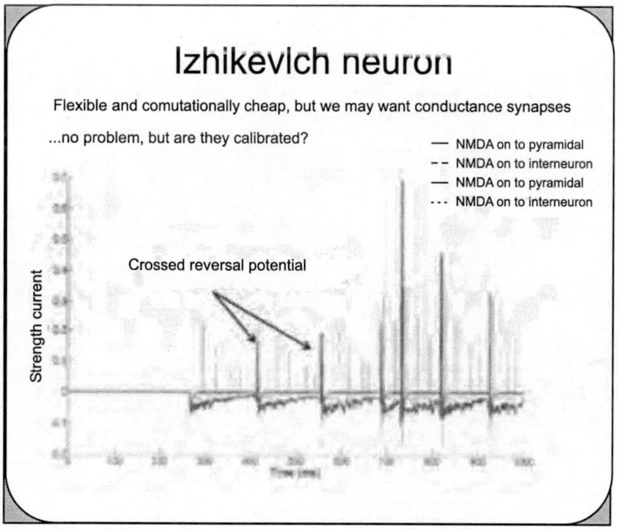
Izhikevich neuron
Flexible and comutationally cheap, but we may want conductance synapses
...no problem, but are they calibrated?
NMDA on to pyramidal
NMDA on to interneuron
NMDA on to pyramidal
NMDA on to interneuron
Crossed reversal potential
Strength current

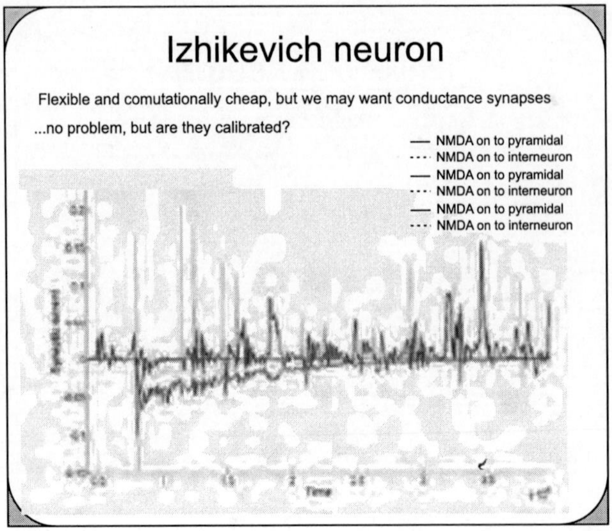
Izhikevich neuron
Flexible and comutationally cheap, but we may want conductance synapses
...no problem, but are they calibrated?
NMDA on to pyramidal
NMDA on to interneuron
NMDA on to pyramidal
NMDA on to interneuron
NMDA on to pyramidal
NMDA on to interneuron

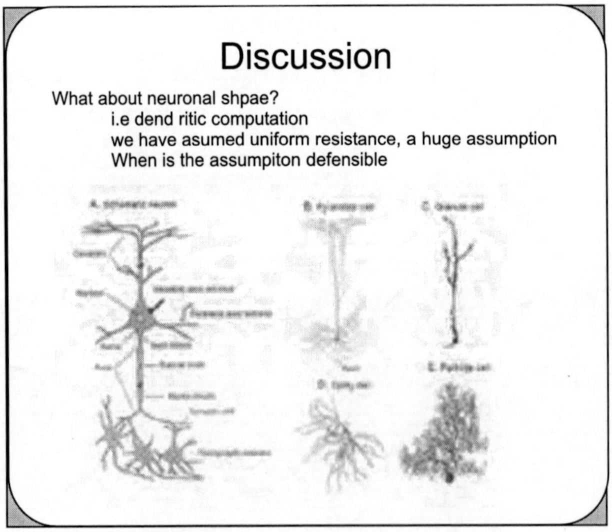
Discussion
What about neuronal shpae?
i.e dend ritic computation
we have asumed uniform resistance, a huge assumption
When is the assumpiton defensible

Discussion

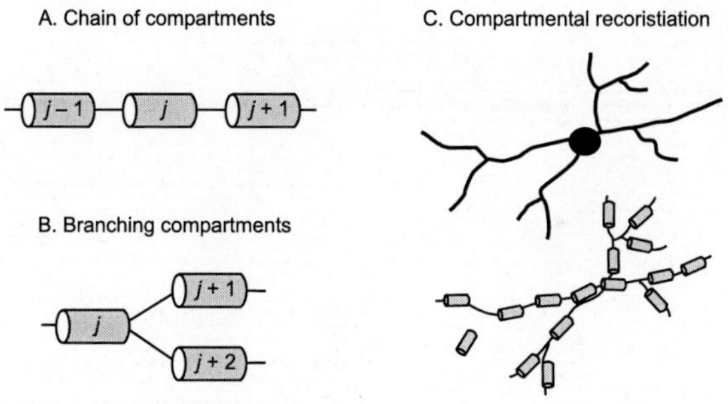

Transparting, fundamental of computational Neuroscience, 2nd ed.

References

E.R. Kandel, J.H.Schwartz and T.M. Jessell, Principles of Neural Science, 4th ed.

T. Trappenbnerg, Fundamental of Computation Neuroscience, 2nd ed.
http://torch.cs.dal.ca/"tt/fun......................

W. Gerstner and W.Kistler, Spiking Neuron Models
http://icww.epfi.ch/ gesstne/BUCH.html

P.Dayan and L.F. Abbott, Theoretical Neuroscience

http://www.izhikevich.com

I

INDEX

L

M

N

O

P

R

S

❑❑❑

Win Prizes!

ATTENTION READERS

We request you for your frank assessment regarding some of the aspects of the book given as under:

NEURAL NETWORKS AND FUZZY LOGIC

By

Dr. R.P.Das and L. Sreedhar

Please fill up the given space in neat capital letters.

(i) What topics of your syllabus that are important from you examination point of view are not covered in the book ?

(ii) What are the chapters wherein the treatment of the subject matter is not systematic or organised or updated ?

(iii) Have you come across misprints/mistakes/factual inaccuracies in the book? Please specify the chapters and the page numbers.

(iv) What other books on the same subject have you found/heard better that the present book? Please specify in terms of price and quality.

(v) Further suggestions and comments for the improvement of the book:

(PS: If need be please attach a separate sheet to write your views.)

Other Details :

(i) Who recommended you the book ? (Please tick in the box near the option relevant to you.)

☐ Teachers ☐ Friends ☐ Bookseller

(ii) Name of the recommending teacher, his designation and address.

....

....

....

....

(iii) Name and address of the bookseller you purchased the book form :

....

....

....

....

(iv) Name and address of your institution (Please mention the University or Board, as the case may be)

....

....

....

....

(v) Your name and complete postal address :

....

....

....

....

(vi) Course you are enrolled in :

....

....

....

....

The best assessment will be awarded each month. The award will be in the form of our publications as decided by the Editorial Board.

Please mail the filled up coupon at your earliest to:

S.K. Kataria & Sons®

4885/109, Prakash Mahal, Dr. Subhash Bhargav Lane,
Daryaganj, New Delhi-110002 (India)
Phone : +91-11-43551243; Mobile: +91-9871775858, 7678425209
e-mail: katariabook@gmail.com, katariabooks@yahoo.com
www.skkatariaandsons.com